WELCOME TO THE MOON

BY MARIANNE J. DYSON AND LINDSEY COUSINS
WITH FOREWORD BY ANDREW ALDRIN
COVER ART BY ALAN BEAN

Copyright ©2019 by Marianne J. Dyson and Aldrin Family Foundation. All rights reserved. No part of this publication may be reproduced, or stored in a retrieval system, or transmitted in any form or by any means, electronic, mechanical, photocopying, recording, or otherwise, without written permission of the Foundation. For rights permission inquiries, please contact the author via www.MarianneDyson.com or write to:
Aldrin Family Foundation
1050 W. NASA Boulevard, Room 109
Melbourne, FL 32901
https://aldrinfoundation.org

Cover design by Lindsey Cousins with art by Alan Bean used with permission
Trade edition ISBN: 978-0-578-47015-3
Printed in the United States of America

Acknowledgements

Marianne Dyson wishes to thank Jim Christensen, Christina Korp, and Andrew Aldrin for the opportunity to inspire a new generation of space explorers. Thank you to Lindsey Cousins for partnering with me to create this book: I couldn't have done it without you! A special thank you to Peter Eckhart for serving as a technical reviewer, and most of all to my family for encouragement and support during preparation of this book.

Photo and Image Credits: Front Cover and spine: ©Alan Bean (art) reprinted with permission, design by L. Cousins; Back cover: L. Cousins. 4: NASA/Goddard/Arizona State University; 5: courtesy A. Aldrin; 6: NASA/JPL/USGS; 7: NASA/JPL-Caltech; 9: NASA as15-87-11835; 10: courtesy Frederick Steiling; 11: NASA; 12: NASA/JSC S80-37406; 13: NASA (Earth left), NASA Goddard (Moon), NASA/GSFC Scientific Visualization Studio (Earth right), drawing by authors; 14: NASA/D. Scott (retroreflector), courtesy Frank Armstrong/UT-Austin (laser beam); 15: NASA Goddard (Moon), NASA/GSFC Scientific Visualization Studio (Earth), drawing by authors; 16: NASA S71-41810 (Apollo 15); 17: NASA ISS038-E-042112, 18: NASA/GSFC/Arizona State University (Near), NASA/JPL (West); 19: NASA/GSFC/Arizona State University, NASA as11-44-6667; 20: NASA; 22: NASA; 23: NASA (Luna 2, Pioneer 4, Ranger 7); 24: NASA; 25: NASA Ranger 7 B001, NASA Ranger 8 A060, NASA Ranger 9 B001, A035, A060, A070; 26: NASA Goddard/Arizona State University; 27: NASA; 28: Mark Wilson (turtle), NASA 67-HC-21 (Apollo 1 crew); 29: NASA/B. Aldrin; 30: NASA; 31: NASA; 32: NASA/N. Armstrong; 33: NASA as11-40-5850; 34: ©Alan Bean (top image, "Fun is Where You Find It") reprinted with permission, NASA/A. Bean (Surveyor/Conrad); 35: NASA S70-035013; 36: NASA 1971 film shot, reprinted with permission of Don Eyles & Sam Drake; 37: NASA as14-64-9118; 38: NASA (Genesis rock), NASA as15-84-11250 (Silver Spur); 39: NASA/D. Scott; 40: NASA s72-37155; 41: NASA as16-113-18347 ; 42: NASA, NASA/E. Cernan/Tom Dahl (Shorty Crater); 43: NASA; 44: NASA/GSFC/Arizona State University; 45: Luna 24, CC BY SA 3 (background removed); 46: NASA AS14-68-9452 (hammer). NASA as16-106-17413 (Young); 47: NASA/GSFC Scientific Visualization Studio; 48: Ernie Wright, NASA Scientific Visualization Studio; 49: NASA; 50: NASA (Moon), USGS (Shoemaker); 51: LPI/Paul Spudis (3 images Moscow Sea); 52: JAXA/NHK; 53: UCLA/NASA/JPL/Goddard; 54: UCLA/NASA/JPL/Goddard; 55: NASA/JPL-Caltech/IPGP; 56: P.H. Shultz & D.A. Crawford, "SPA Impact Origin," LPSC 46, 2015; 57: NASA; 58-59: China National Space Administration (CNSA) and Chinese Lunar Exploration Program (CLEP); 60: Huáng zhú shuǐ shēng- Own work, CC BY SA 4.0; 62: Openclipart.org (Earth), drawing by authors; 63: NASA; 64: NASA as11-69-21294; 65: U.S. Air Force/Joe Davila (Delta IV), SpaceX (middle), Proton M (right), CC BY-SA 2.0; 66: Northrop Grumman; 67: NASA s69_39958; 68: ©M. Dyson; 69: Lockheed Martin; 71: ©M. Dyson; 72: RSC Energia (Tereshkova), NASA (Ride), NASA (Ansari), CNS (Yang); 73: Clem Tillier; Earth graphic based on NASA image of Earth seen from Apollo 17, CC BY SA 2.5; 74: NASA; 75: ©M. Dyson; 76: NASA exp52-iss052e019970; 77: NASA iss050e029710; 78: NASA; 80: NASA ISS-007e11796; 81: NASA/Goddard/Arizona State University; 82: LRO/NAC Images with LOLA DEM data/ Jon Meyer 2011; 83: NASA ISS021-03E0778; 85: NASA as16-113-18339; 87: ©M. Dyson; 88: NASA; 91: ©Alex Aurichio, 2009, reprinted with permission.

TABLE OF CONTENTS

Foreword by Andy Aldrin 5

CHAPTER 1: MEET THE MOON 6
    Congratulations, It's a Moon! 7
    The Changing Face of the Moon. 9
    Sidebar: Look Up for Directions 12
    Diagram: Opposite Phases 13
    The Pull of the Tides 15
    Escaping Earth 16
    Sidebar: No Zero G 17
    Lunar Tour 18
    Lunar Orbit. 19
    Descent to the Surface. 21

CHAPTER 2: HISTORY OF EXPLORATION 22
    Probing the Moon 23
    Timeline 1959-65. 23
    First Landings on the Moon 26
    Preparing for Human Missions 28
    Sidebar: How Hard is it? 29
    Earthrise from the Moon 30
    Timeline 1966-69 31
    Apollo 11 Tranquility Base. 32
    Apollo 12 The Artist's Moon. 34
    Apollo 13 A Successful Failure 35
    Apollo 14 Pushing the Limits 36
    Timeline 1970-76 37
    Apollo 15 Training Pays Off 38
    Apollo 16 A History of Impacts. 40
    Apollo 17 A Fountain of Knowledge . . . 42
    Soviet Lunar Exploration 44
    History Told by Apollo Rocks 46
    Activity: Connect the Apollo Dots. 47

CHAPTER 3: NEW VIEWS OF THE MOON 48
    Hidden in Shadow 49
    Sidebar: Mapping the Minerals. 51
    The International Moon 52
    Deep Secrets. 55
    Glowing Neon Signs. 57
    Back to the Surface 58

CHAPTER 4: HUMAN RETURN . . . 60
    Staging the Launch 61
    To the Moon and Back Apollo Style . . . 62
    Moving Out of Earth Orbit 65
    Sidebar: Gassing Up 67
    New Crew Modules 68
    Facing the Fire 71
    Sidebar: First Woman on the Moon . . . 72
    Staying on the Moon 73

CHAPTER 5: LONG-TERM STAY 74
    Safety First 75
    Power from the Sun. 76
    Going Nuclear 78
    Water: Fuel for Life 79
    The Lunar Underground. 81
    Greening the Moon 83
    Flying to Stay Healthy 84
    Working on the Moon 86

CHAPTER 6: HEADING OUT 88
    The Value of the Moon 89
    MoonBase One 90

Units Conversion/Moon Facts 92
Glossary 93
Index 95

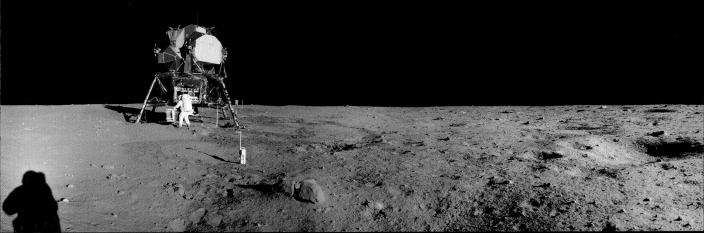

People orbiting the Moon will see Earthrise about 12 times a day just like NASA's Lunar Reconnaissance Orbiter as it flew about 83 miles above the Moon's north pole in 2015.

FOREWORD

Fifty years ago, my father, Buzz Aldrin, walked on the Moon during Apollo 11. It was one of the most remarkable feats of humanity. Indeed, it is important to remember that it was a feat of humanity. The plaque on the Apollo 11 lander reads, "We came in peace for all mankind." During the next three years, the United States sent six more Apollo missions to the Moon. Since that time, no human has set foot on the Moon. That is about to change. Welcome back to the Moon!

Yes, humanity is going back to the Moon. But this time the United States will not go alone. As you will discover in this book, Russia, China, Japan, India, Israel, and European nations already have spacecraft studying the Moon. Private companies have plans to take people to the Moon and to set up new space industries. And this time, you too will have an opportunity to participate, not only as astronauts and scientists, but as business executives and workers. Soon, you may even go to the Moon for vacation!

The authors and I thank you for reading this book and encourage you to share it with others. We hope after learning how we got to the Moon 50 years ago, what we have done since then, and the economic and scientific benefits the Moon has to offer today, you too will want to play an active part in the amazing return to the Moon. By working together, humanity can return to the Moon, this time to stay!

Andrew Aldrin
President
Aldrin Family Foundation

CHAPTER 1
MEET THE MOON

Welcome Lunar Pioneers! This briefing book is packed full of knowledge about the Moon that we hope will enhance your lunar stay, be it for vacation, work, or as your new home. First up is understanding the forces that shaped the Moon into the world it is today.

A giant impact formed the Moon.

CONGRATULATIONS, IT'S A MOON!

Most scientists agree that our solar system formed from a nebula, a spinning ball of gas and dust, about 4.6 billion years ago. The heat of the newborn Sun blew the outer layers of the nebula away, like steam from a boiling pot. The farthest gases chilled into the outer gas giants. Heavier elements such as iron didn't rise as far and cooled to form the inner rocky worlds. Thus, each world has a different composition depending how far from the Sun it was when it formed. There's more metallic iron on Mercury than there is farther away on Mars. A meteorite from the asteroid belt has a mineral mix different from one from Mars.

Moon rocks contain the same minerals as Earth rocks. Therefore scientists think the Moon and Earth formed at the same distance from the Sun.

WELCOME TO THE MOON

However, the Moon's small core shows that it did not form like Earth. The core of Mercury is about 60 percent of the planet's mass whereas the core of Earth is about 30 percent. The core of Mars is only about 10 percent of its mass. But the Moon's core is only one percent of its mass. How can the Moon have the same mineral mix as Earth but lack a big iron core?

The GIANT IMPACT THEORY currently offers the best way to explain these observations. It goes like this: While Earth was still hot and fresh from its own formation, a world, about half the size of Earth, slammed into it. The impact was off-center, so the impact spun up the early Earth and flung molten material into space. This material cooled and condensed into a ring around the Earth from which the Moon formed. As the Moon cooled and shrank, pressure inside caused lava to erupt onto the surface, flowing into low places. Over billions of years, more impacts and eruptions altered the Moon's composition and appearance.

The Giant Impact Theory explains why the Moon has the same composition as Earth's crust and yet has a small iron core: it is made out of the MANTLE of Earth.

However, questions remain unanswered such as when exactly the impact happened, if there was more than one formation impact, and how long it took for the Moon to cool. Rocks brought back from the APOLLO missions have been key to unraveling the answers. New sample returns by robots and humans will provide a more complete understanding and help locate needed resources.

THE CHANGING FACE OF THE MOON

The Moon started out so hot that its surface was liquid rock called MAGMA. The heavier minerals in the magma sank and became the Moon's core. The surface cooled and formed a crust made of a lightweight rock called ANORTHOSITE. This happened within about sixty million years.

The early solar system was like a giant pinball parlor. Smaller bodies were in a gravitational tug of war with the newborn planets and the Sun.

Mountains on the Moon, like Apollo 15's Mt. Hadley shown here, are round instead of pointy because they are buried under a thick layer of regolith.

WELCOME TO THE MOON

The baby Moon got pounded. The force of the biggest impacts raised rings of mountains, some taller than any on Earth. The impacts melted and shattered the crust and hurled debris that "rained" down all over the Moon. This ground up rock and dust is called REGOLITH.

As the Moon cooled, it shrank and "squeezed" the hot liquid under the crust so that it pushed up and burst onto the surface. The lava that came out of these lunar volcanoes was about the consistency of motor oil.

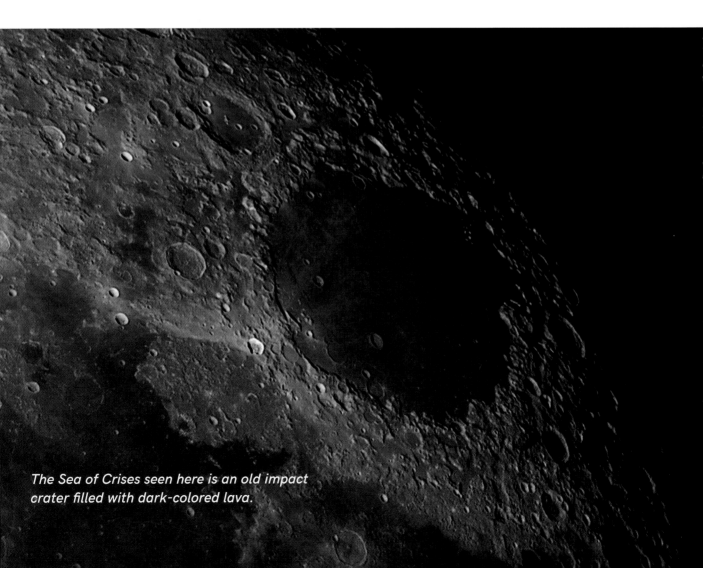

The Sea of Crises seen here is an old impact crater filled with dark-colored lava.

HOW OLD IS THE MOON?
Inside some Moon rocks (Apollo 14 sample shown) are light-colored bits of the original lunar crust. Some bits include zircon crystals that contain uranium that decays into lead at a known rate. By comparing the amount of uranium and lead, scientists can determine the age of the rock, and hence the Moon. The oldest rock, from Apollo 14, is 4.51 billion years old.

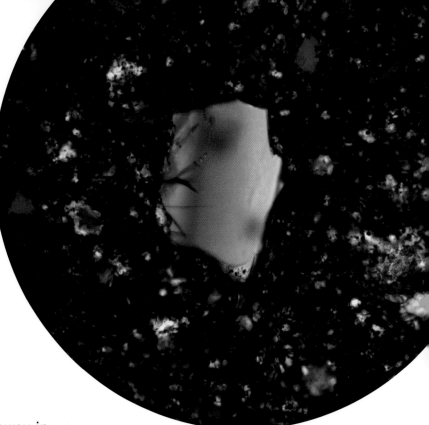

The lava came out and flowed away instead of building up large volcanic mountains.

The lava flowed into the low places and hardened into a dark-colored rock called BASALT. The largest pools of lava are called MARIA (pronounced "mah'-ree-ah"), the Latin word for seas. The maria are round because the craters were round, not because they are the tops of a giant volcanoes.

Impacts continued throughout the Moon's history. Some impacts, such as the ones that produced Tycho and Aristarchus, were powerful enough to punch through the dark layers of basalt, exposing and spraying out rays of the whiter crust underneath. The heat of these impacts produced a kind of rock called BRECCIA (pronounced "brech-ee-uh"). Breccias are conglomerates of other rocks such as basalts with bits of regolith and the impacting body melted and stuck together.

WELCOME TO THE MOON

LOOK UP FOR DIRECTIONS

Earth does not rise and set as seen from the NEAR SIDE surface. Instead it holds steady like a beacon in the black sky. At Central Bay (marked with red cross on page 13), Earth is directly overhead: 90 degrees up from the horizon in all directions.

When moving away from that location, Earth appears to slide down toward the horizon opposite to the direction of travel. Earth sinks below the horizon on the FAR SIDE. The boundary between the near and far sides are the only places on the surface where the Earth rises and sets. This "peek-a-boo" or bobbing of Earth north and south is the result of the tilt of the Moon's orbit.

People on the far side can't see Earth, but they can use the Sun for directions. (The entire far side is only dark when the Moon is full!) Just like on Earth, the Sun rises in the east and sets in the west. At the north pole, the Sun moves counterclockwise around the horizon. At the south pole, the Sun moves clockwise.

The stars can also provide directions. However, because the Moon's spin axis is not pointed in the same direction as Earth's axis, the North Star (Polaris) is not above the Moon's north pole. Instead, as the Moon makes one rotation every 27.3 days, the constellations appear to rotate around a spot in the sky about half way between Polaris and the brightest star in the constellation Draco.

Stars are always visible from either side of the Moon, but hard to see during DAYSPAN. They are also hard to see during NIGHTSPAN on the near side because of Earthshine. A full Earth is as bright as a 60-Watt bulb about seven feet away.

To see stars, pioneers can dark adapt their eyes by avoiding bright light for about 30 minutes. Apollo astronauts did this and then used a telescope inside their lander to determine precisely where they were. About every six months, the full Moon passes through Earth's shadow. (Usually it is above or below the shadow.) During the ECLIPSE, the Earth's atmosphere lights up in a spectacular ring—visible on the near side and by Apollo 12 on their way home.

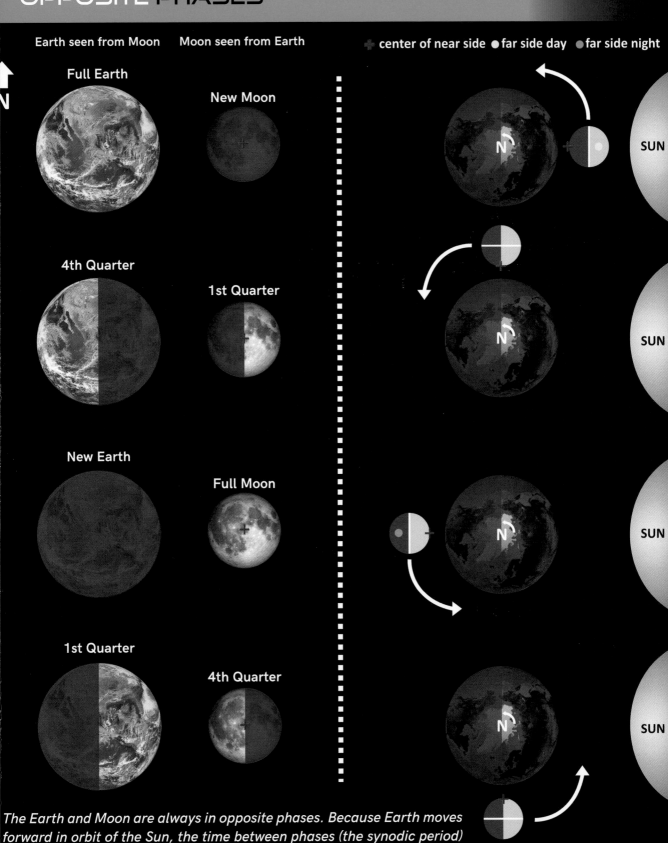

The Apollo 15 laser ranging retroreflector lower rows show the reflection of the black sky.

A laser beam is projected from McDonald Observatory in Texas to bounce off the reflector and back, measuring the distance to the Moon. The beam is only visible while in the Earth's atmosphere.

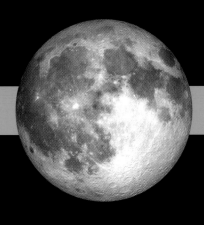

THE PULL OF THE TIDES

Because the Moon rotates once each revolution, only the Near Side is visible from Earth.

The Moon formed much closer to Earth than it is now. Scientists estimate it was only 14,000 miles away compared to about 240,000 miles now. When the Moon was closer, its GRAVITY created tides of thousands of feet on Earth. Likewise Earth caused huge tides on the molten Moon. These forces slowed the spin of both worlds dramatically at first, and more slowly now. The Earth slowed from about a five-hour day to 24 hours per day. The Moon's spin slowed to once every 27.3 days, the same period as one revolution. Because of this SYNCHRONOUS ROTATION, only the NEAR SIDE is visible from Earth.

The same force that is slowing the Earth's spin is pushing the Moon outward. The tidal bulge on Earth created by the Moon is not directly below the Moon. Instead, it is shifted "forward" because moves outward, its gravitational tug gets weaker, the bulge gets smaller, and the increase in velocity is less. The Moon's retreat has slowed to about 1.6" per year. At this rate, the Moon may be de-

ESCAPING EARTH

Launching from Earth is fast and furious! From the launch pad to space takes less than ten minutes! To stay in space (and not fall back down like a ball tossed in the air), the rocket must go fast enough to balance the force of Earth's gravity.

Once that speed is reached, the engines shut down, and the rocket "coasts" in LOW EARTH ORBIT (LEO, pronounced "lee-oh"). Low Earth orbits can be up to 1200 miles high, but altitudes of around 200 miles are typical for human flights.

The "layover" in Earth orbit offers passengers a gorgeous view while giving the crew time to check spacecraft systems before committing to the Moon.

Gravity is often described as the shape of space. Each body of mass makes a depression or well around it like balls sitting on a foam mat. To

The first two stages of the Saturn V booster (Apollo 15 launch shown) lifted the Apollo command and service modules to Earth orbit. The third stage later boosted them to the escape velocity required to reach the Moon.

NO ZERO G

Despite what you may have heard, there **is** gravity in space! The force of gravity depends on the mass of the body divided by the square of the distance from its center. Earth's gravity, abbreviated g, equals 1 at the surface of Earth.

In orbit, gravity is less because the distance has increased. But it is NOT zero. Imagine a bathroom scale on a tower 200 miles tall. Because gravity is measured from the center of the Earth (4000 miles), someone standing on that scale is not that much farther away. Gravity is actually 90 percent of what it is on the surface. And if that person stepped off the tower, they would fall quickly to Earth.

Spacecraft in orbit are falling, too. They don't hit the ground because, like a ball on a string being twirled in a circle, they are moving fast enough to balance gravity's pull. Spacecraft must reach and maintain a velocity of about 5 miles per second to keep from falling back to Earth. So if there is gravity in space, why are astronauts weightless?

Astronauts are weightless because they are in FREEFALL. The spacecraft and everything inside appear to be floating because they are all falling together like people who skydive from an airplane. If they push someone beside them, they move apart but continue falling. Astronauts float AS IF there were no gravity. Unfortunately, most people drop the AS IF part of the sentence, leading to **mass** confusion. (Sorry for the pun!)

Spacecraft must slow down to fall back to Earth. The tricky part is to slow down gradually to avoid burning up from friction with the atmosphere.

On the way to the Moon, passengers will experience freefall and enjoy a fantastic view of Earth.

WELCOME TO THE MOON

keep from falling back into the "well" of a body, a spacecraft must go fast enough to climb out of the bowl. This speed is called the ESCAPE VELOCITY. When Mission Control gives the "go" for TRANSLUNAR INJECTION (TLI), the engines fire long enough to reach this speed.

After about two days of climbing "uphill," the spacecraft reaches the intersection of the Earth's gravity well with the Moon's shallower well.

Even though the ship is moving incredibly fast, Apollo astronauts said they didn't feel the speed. But they did notice that during the trip, Earth went from filling the window to small enough to blot out with a thumb!

LUNAR TOUR

Orbiting the Moon counterclockwise (east to west) reveals four views of the Moon—near, west, far, and east sides shown in sequence. Note that to see the entire surface in daylight would take 28 days.

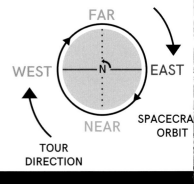

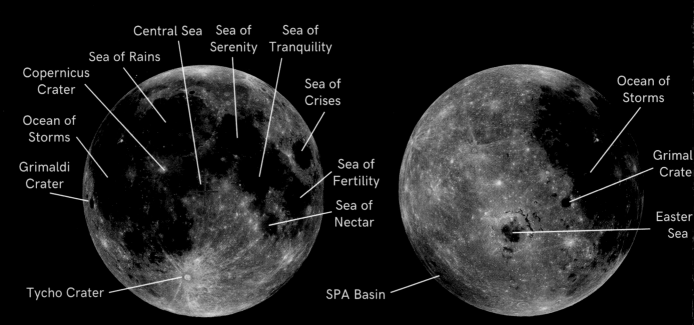

NEAR Side. *The Sea of Crises marks the eastern limb. The east also features the "upside down" rabbit with Tranquility as its "head." Central Bay is 0 latitude/0 longitude (red crosshairs). The Sea of Rains (left "eye") and Ocean of Storms dominate the west. Tycho is the white-rayed crater in the south. Dark Grimaldi Crater marks the western limb.*

WEST Side. *The Ocean of Storms and Grimaldi Crater are still visible on the right. The Eastern Sea is the dark ringed basin in the center. The left side is the far side not visible from Earth. The "shadow" of the South Pole Aitkin (SPA) basin is visible in the lower left.*

LUNAR ORBIT

As the ship nears the Moon, the pull of its gravity gets stronger and stronger. The ship speeds up just like a ball rolling downhill. Unless the ship slows itself down, it will hit the surface at high speed.

High speed impacts can be useful for science, but humans need a gentle landing. Also, people may want to leave their ride home in orbit instead of having to use energy (fuel) to slow it down. The Apollo COMMAND AND SERVICE MODULES with one astronaut onboard were left in orbit while two men went to the surface in a LUNAR MODULE (LM, pronounced "lem").

It takes about a third as much "DELTA V" (change in velocity) and thus less fuel for LUNAR ORBIT INSERTION (LOI) as it does to go directly to the surface.

The view from lunar orbit is spectacular! As spacecraft orbit counterclockwise, the mysterious far side will be unveiled—how much depending on the phase of the Moon.

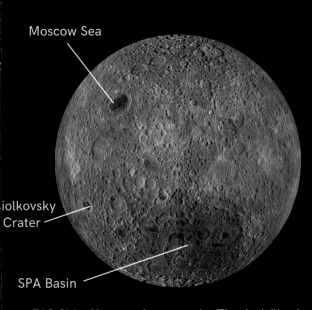

FAR Side. Note, no large maria. The dark "bruise" in the south is the SPA basin. The Moscow Sea is on the upper left. Tsiolkovsky Crater is near the eastern limb.

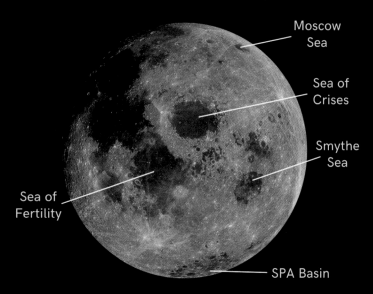

EAST Side. The Moscow Sea and SPA are still visible. The Sea of Crises is in the center with Smythe Sea to the right. This photo was taken by the Apollo 11 crew. It is often mislabeled as the near side but (the right) half is not visible from Earth.

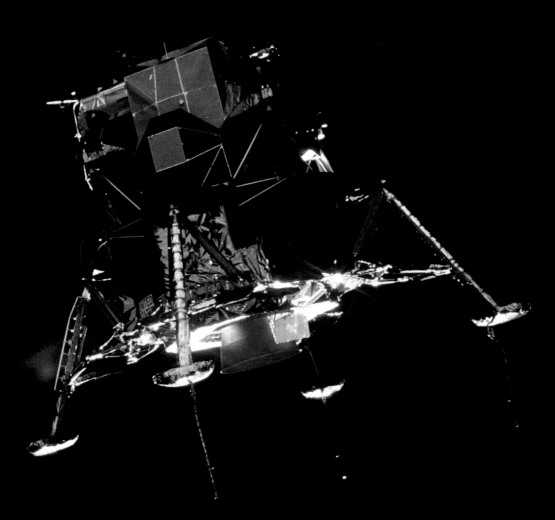

Apollo 11 descends to the Moon. Note the thin tubes hanging below three of the footpads. (The leg with the ladder didn't have one to prevent it from puncturing a space suit during egress.) When one of these probes hit the surface, the contact light lit. The engine was shut off before touchdown to prevent the exhaust from sand-blasting the lander.

THE DESCENT TO THE SURFACE

During the Apollo Program, twelve men made the trip to the lunar surface. The crew of Apollo 11 were the first to land.

The Eagle was packed with pipes and wires and sprayed-on gray insulation. The men sealed the hatches. Then Mike Collins, the Command Module pilot, pulled away in Columbia with a snapping thump. Neil Armstrong and Buzz Aldrin "stood" shoulder to shoulder in spacesuits and bubble helmets, tethered to the deck by elastic cords. The Moon seemed to rotate past their windows until it hung "above" their heads. "The Eagle has wings," Neil told Houston.

They dropped down to 8 miles above the Moon. Buzz reported that the engine firing was so smooth, he didn't feel it. The LM flew "feet" forward with the windows facing the surface. They dropped lower and throttled up. They felt a slow sagging in their knees from the Moon's gravity. Neil grinned at Buzz through his helmet.

As they got closer, the Moon changed from beige to gray. The radio hissed and crackled when Houston said they were "go."

An alarm sounded. Mission Control said to keep going.

They dropped to about 4,000 feet, then 2,000 feet. Another alarm flashed. The computer was overloaded. But Mission Control said to keep going.

At 700 feet, yet another alarm flashed! Neil punched PROCEED on the keyboard.

Neil saw a boulder field out his window. He slowed the descent. They skimmed forward. Buzz called out the altitude. Neil scooted past the boulders.

The low-fuel light blinked. They had only 60 seconds of fuel left. Then 30 seconds. They entered the "dead man's" zone. If they ran out of fuel, they would crash before the ascent engine could engage. "Forward. Forward. Good. Forty feet," Buzz reported.

The footpad probe touched the surface. "Contact light," Buzz called. He said the view outside was as stark as he'd ever imagined. A mile away, the horizon curved into blackness.

"Houston," Neil called, "Tranquility Base here. The Eagle has landed."

CHAPTER 2
HISTORY OF EXPLORATION

Many space missions preceeded the first human landing on the Moon. These missions were vital to developing the skills and information needed for the first and future explorations of the Moon to be successful.

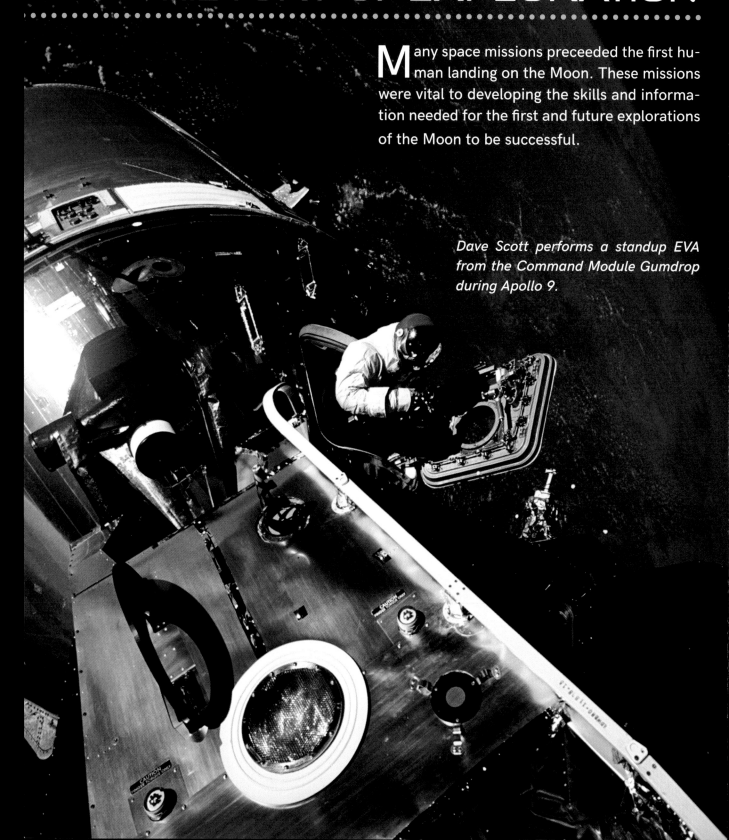

Dave Scott performs a standup EVA from the Command Module Gumdrop during Apollo 9.

PROBING THE MOON

Humans have enjoyed observing the Moon throughout history. Starting with Galileo (1564-1642) in the 1600s, they were able to see more details using telescopes. By the early 20th century, scientists had mapped all the major features and debated the origin of craters on the Moon's surface—were they the tops of volcanoes or the result of impacts? (Studies of high speed impacts and Meteor Crater in Arizona pointed to an impact origin confirmed later by Apollo rocks.)

The Soviet Union, now Russia, launched the first satellite, Sputnik, in 1957. The United States launched Explorer 1, in 1958. Soon after these events, scientists had the opportunity to study the Moon in a brand new way: with spacecraft.

On January 2, 1959, the Russian Luna 1 became the first spacecraft to fly by the Moon. The United States launched its first flyby, Pioneer 4, in March of that year. Luna 2 became the first manmade object to reach the surface when it crashed on the near side on September 12. Then Luna 3 provided the first ever look at the far side on October 4. The world was amazed to see that the far side lacks the large dark maria seen on the near side.

1959
Jan: Luna 1*
March: Pioneer 4
top image
Sept: Luna 2*
second image
Oct: Luna 3*

1961
May: JFK sets goal
Aug: Ranger 1
Nov: Ranger 2

1962
Jan: Ranger 3
April: Ranger 4
Oct: Ranger 5

1963
April: Luna 4*

1964
Feb: Ranger 6
July: Ranger 7
pictured right

1965
Feb: Ranger 8
March: Ranger 9
May: Luna 5*
June: Luna 6*
July: Zond 3*
Oct: Luna 7*
Dec: Luna 8*

*Russian missions

WELCOME TO THE MOON

In 1961, President John F. Kennedy announces goal of putting a man on the Moon.

Learning about the Moon took on a new urgency in 1961. On April 12, the Russians launched the first human, Yuri Gagarin, into space. This feat by a communist regime was viewed as a display of military strength: if they could put a man in space, they could beat any foe in battle anywhere.

To counter this threat, the American President John F. Kennedy, chose to do something even more difficult: put a man on the Moon by the end of 1969. New technology and science were required to win this "space race."

Landing, even crashing on purpose on the lunar surface, is hard. The Russian's Luna 4 failed in 1963 as did the first three American Rangers. Ranger 4 earned the dubious honor of being the first manmade object to impact the far side in 1962. Ranger 5 lost power and missed the Moon altogether, and Ranger 6 crashed in the Sea of Tranquility in 1964 without returning any images.

Success was finally achieved on July 31, 1964. The American Ranger 7 sent back more than four thousand images before impacting (on purpose) in the Known Sea.

Ranger 8 impacted in the Sea of Tranquility in February 1965. Its images showed that areas that looked smooth from Earth were actually pitted with craters.

Ranger 9 sent the first television images from the Moon and impacted in the crater Alphonsus in the lunar highlands on March 24. Data from this mission led to the discovery that the Moon's center of mass is offset from its geometric center.

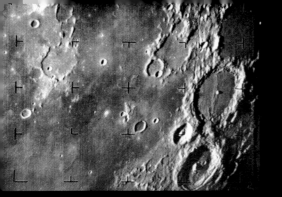
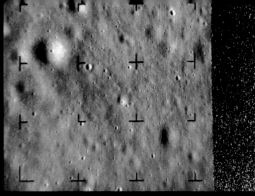

Ranger 7 was the first American spacecraft to return images of the surface of the Moon. This image shows the Known Sea, near the Apollo 14 landing site on the near side.

Ranger 8 imaged an area about 37 miles from the Apollo 11 landing site in the Sea of Tranquility. The right side is missing because the spacecraft crashed before completing transmission.

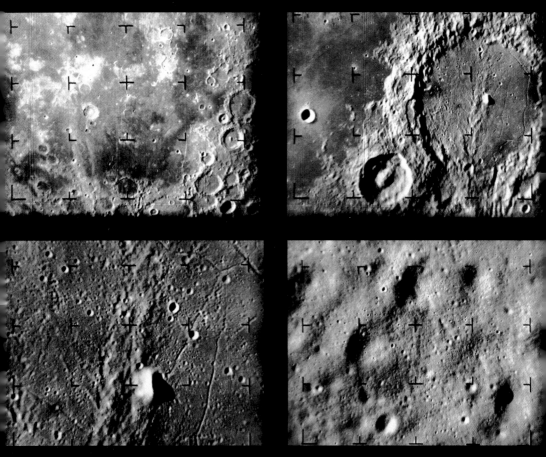

These four images from Ranger 9 show the view of Alphonsus Crater from 1553 miles, then 275 miles, then 85 miles, and finally from 5 miles, taken 3 seconds before it crashed to the right of the central peak in the previous image.

Surveyor 7 captured an image of Tycho Crater in 1968. That year's hit movie, 2001: A Space Odyssey, placed the alien obelisk here. The central peak is 6,500 feet tall.

FIRST LANDINGS ON THE MOON

The Russians continued to perfect their spacecraft systems. In 1965, Luna 5 became their second spacecraft to hit the Moon. It impacted in the Sea of Clouds, though not where they had originally intended. Luna 6 failed, and control was lost of both Luna 7 and 8. Both crashed in the Ocean of Storms.

But Luna 9 was an amazing success. On February 3, 1966, it descended to the lunar surface. At an altitude of 46 miles, it

The Soviet Luna 9 weighed about as much as a mid-sized car (3390 lbs) and stood nine feet tall. The central sphere was about two feet wide and held the radio, programs, batteries, and thermal control systems. Four antenna opened after landing.

inflated airbags and fired its landing rockets. When it was 16 feet above the surface, the engines shut down as a sensor contacted the ground. The landing capsule, in its airbag cocoon, was ejected and bounced several times before coming to rest in the Ocean of Storms west of Reiner crater. The four petals of its spherical body opened to allow the antenna to spring out. Television cameras revolved and tilted, providing a panoramic view of the nearby rocks with the Sun poking just 3° above the black horizon. The stunning images revealed the spacecraft had landed in a small (82-foot) crater on a slope of about 15° that increased to 22.5° as the regolith settled, and the craft slid down. This historic spacecraft has sat silently on the lunar surface since its batteries ran out on February 6, 1966.

In June, the American Surveyor 1 landed in the Ocean of Storms (to the west of Apollo 12's landing site). It fired retrorockets to slow down from 6000 mph to 3 mph and landed on legs, no airbags or bouncing involved. It stood about 10 feet tall and weighed about 650 lbs at landing. The three footpads (folded for launch) extended out 14 feet from the center. It used a large antenna dish mounted on top to transmit television and other data to Earth. This hardy spacecraft was equipped with solar cells that recharged the batteries, allowing it to survive the two-week lunar nightspan. The next Surveyor crashed near Copernicus Crater in September.

In 1967, Surveyor 3 landed in the Ocean of Storms, 4 crashed in Central Bay, 5 landed in the Sea of Tranquility, and 6 landed in Central Bay. The final Surveyor landed far south of the lunar equator in Tycho Crater in January 1968. The Surveyor spacecraft successfully validated the landing design planned for Apollo.

The Russians launched three Luna orbiters in 1966 and one more lander, Luna 13. Similar to Luna 9, it landed in the Ocean of Storms. But Luna 13 had some fancy new experiments, one to measure the radiation emitted from the soil, and another to probe the hardness of the surface. Two more Luna orbiters launched, one in 1968, and another in 1969. Their accomplishments were soon overshadowed by Apollo.

PREPARING FOR HUMAN MISSIONS

Desert turtles were the first living creatures to fly to the Moon, on Zond 5 in September 1968.

The Americans planned to send three men to the Moon and land two of them on the surface. The Russians planned to send two and land one cosmonaut. Rather than using a docking tunnel, the cosmonaut would spacewalk to and from the lunar lander. Both plans were fraught with challenges.

The original design of the American command module had hatches (doors) that opened only from the outside. During a ground test in January 1967, a fire broke out inside the capsule. The Apollo 1 crew of Gus Grissom, Ed White, and Roger Chaffee were killed. This tragedy drove a redesign of the capsule and a focus on crew safety.

The Russians needed to reduce the weight of their Soyuz capsule to use it for lunar missions. They removed the habitation module and added a better HEAT SHIELD and computer. The two-man cap-sule was renamed L1, but officially known as Zond (which means probe in Russian). The first unmanned test was on Kosmos 146 in March 1967. Two weeks later, they too experienced a tragic loss of life. Cosmonaut Vladimir Komarov was killed during the entry of Soyuz 1.

Then the Zond spacecraft made history by ferrying the first living things to the Moon. In September 1968, two desert tortoises, some fruit flies, meal worms, plants, seeds, and bacteria launched on Zond 5. Additionally, a human mannequin weighing 154 pounds and 5' 8" tall occupied the pilot's seat. The mannequin was loaded with radiation sensors. The turtles were likely chosen because of their ability to go without water for their six-day journey. The capsule

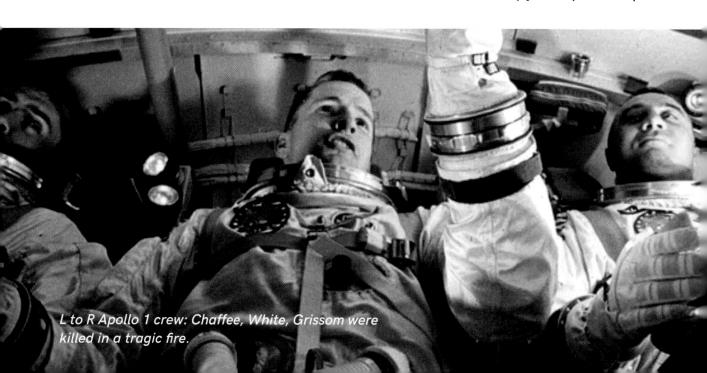

L to R Apollo 1 crew: Chaffee, White, Grissom were killed in a tragic fire.

HOW HARD IS IT?

How strong do the legs of a lunar lander need to be? Will the foot pads sink deep into fluffy powder, break rocks into glasslike shards, or smack into solid stone? To find out, Russian engineers devised an experiment for Luna 13 to test the hardness of the lunar surface.

After bouncing to a stop, Luna 13's airbag deflated, and two booms sprang out from the body of the spacecraft. One boom contained a small solid rocket, pointed down. The rocket shot a TITANIUM cone penetrator (with a diameter of 1.4" into the ground. A pin slid along a groove in the side of the casing to measure how deep it went. The engineers had tested this penetrator on 14 different surfaces on Earth, including dust and concrete, and in a vacuum chamber ahead of time. Depending on the surface material, the penetrator could drill down 2".

On Christmas Eve of 1966, the engineers got the gift of data from the Moon. The penetrator dove in 1.7". The team concluded that the surface was volcanic rock (basalt) covered by a layer of powder.

Spacecraft cameras revealed rocks scattered on the surface. Of 181 rocks counted, most were pebbles. Only three were larger than 4" and all less than 8" in diameter. The experiment gave the engineers confidence they could safely land a cosmonaut on this surface.

Buzz Aldrin tested soil hardness with this boot print. The length of the shadows reveal this print is less than an inch deep.

splashed down in the Indian Ocean after experiencing a g-load that would have killed a crew. But the hardy turtles survived. The Russian news agency announced they had lost only 10 percent of their body weight and showed no loss of appetite.

Zond 6 followed in November. The entry included a complex "skip" maneuver which involved dipping into the atmosphere, then popping out to cool off, and then diving back in. This maneuver allowed the capsule to slow down enough to land in Russia instead of in the ocean. Unfortunately, the parachute failed, and the animals were lost.

EARTHRISE FROM THE MOON

The Apollo 8 crew were the first humans to see the Earth from the distance of the Moon and also to see the far side of the Moon.

Thinking the Russians were about to beat them, NASA managers made a bold decision in 1968. The Saturn V booster was ready. The redesigned command module had been tested in Earth orbit by Apollo 7. But the lunar module was behind schedule. Should they wait for it or go to the Moon without it? They decided to go for the Moon.

Apollo 8 launched on December 21, 1968. Frank Borman, Jim Lovell, and Bill Anders were the first astronauts to ride a Saturn V into space. In the first two and a half minutes, the 12-foot-wide engines of the first two stages burned 54 railroad cars worth of liquid oxygen LIQUID OXYGEN (LOX, pronounced "locks"). After reaching orbit, the third stage fired again, taking them to the Moon. Three days later, they entered lunar orbit while out of contact with Earth. They were the first humans to see the far side. Borman reported, "The Moon is essentially gray. No color. Looks like plaster of Paris."

Four orbits later, they rolled the spacecraft and saw the Earth hovering over the lunar horizon as captured in the famous "Earthrise" photo. On Christmas Eve, their reading from the Bible soothed and inspired millions of listeners after a year of war, assassinations, and political unrest throughout the world.

Even though the Americans had done the first CIRCUMLUNAR flight, the Russians continued to compete to be first to land a human on the surface. The first test of their N-1 booster failed a minute after launch, though the L1 capsule was spared by the ABORT system.

Apollo 9 in March 1969 was the first

human test of the lunar module. James McDivitt, David Scott, and Rusty Schweickart put the vehicle and the new spacesuits through a series of tests in Earth orbit.

On May 18, Tom Stafford, Gene Cernan, and John Young thundered into the sky on Apollo 10. Three days later, while Young stayed in the command module, Stafford and Cernan rode the LM to within 8.7 miles of the lunar surface. They successfully performed a RENDEZVOUS with the command module while out of touch with Mission Control on the far side. They returned safely to Earth a few days later.

Russian hopes of being first to the Moon ended on July 3, 1969. One of the 30 first-stage engines of the N-1 exploded on the launch pad. A fire erupted, the rocket fell over, and the entire launch complex was destroyed. They tried one last-ditch effort to get a lunar rock home before the Americans by launching Luna 15. It crashed in the Sea of Crises on July 21, the day after Apollo 11 landed.

Apollo 7's crew, Wally Schirra, Donn Eisele and Walter Cunningham, successfully tested the redesigned capsule in 1968.

1966
Feb: Luna 9*
March: Luna 10*
June: Surveyor 1
Aug: Lunar Orbiter 1
Aug: Luna 11 orbiter*
Sept: Surveyor 2
Oct: Luna 12 orbiter*
Nov: Lunar Orbiter 2
Dec: Luna 13 lander*

1967
Jan: Apollo 1 fire
Feb: Lunar Orbiter 3
March: Kosmos 146*
April: Kosmos 154*
April: Surveyor 3
April: Soyuz 1*
May: Lunar Orbiter 4
July: Surveyor 4
July: Explorer 35 orbiter
Aug: Lunar Orbiter 5
Sept: Surveyor 5
Oct: Kosmos 186 & 188*
Nov: Surveyor 6
Nov: L-1 failure*

1968
Jan: Surveyor 7
March: Zond 4/L-1*
April: Luna 14 orbiter*
Sept: Zond 5*
Oct: Apollo 7
Nov: Zond 6*
Dec: Apollo 8

1969
March: Apollo 9
May: Apollo 10
July: Luna 15*
Jul: Apollo 11
Aug: Zond 7*
Nov: Apollo 12

*Russian missions

APOLLO 11 TRANQUILITY BASE

On Wednesday, July 16, 1969, Apollo 11 launched on its historic flight. For their trip to the Moon, the top of the LM (Eagle) was joined to the "nose" of the CSM (Columbia). To keep the fuel tanks from getting too hot in the Sun or too cold facing away from the Sun, they rotated in "barbecue" mode. As the craft rotated, Earth appeared to move from one window to another in Columbia.

After three days, they slowed down and entered orbit sixty-five miles above the Moon.

On July 20, 1969 Collins helped Armstrong and Aldrin into the Eagle. About two and a half hours later, they became the first humans to land on another world.

At about 9:30 PM Houston time, Armstrong wriggled out through the hatch. He lowered a small drawbridge that held a TV

How far away is Aldrin from Armstrong (seen in the visor) in this Apollo 11 photo? Because the astronauts stood on a smaller "ball" than Earth, they thought rocks were farther away and bigger than they actually were.

HISTORY OF EXPLORATION

camera. Six hundred million people around the world watched as he backed down the ladder. He touched the Moon's surface and said, "That's one small step for man . . . one giant leap for mankind." Collins missed those famous words. He was on the far side of the Moon, out of earshot.

Aldrin soon followed. "Beautiful! Beautiful!" he said. "Magnificent desolation." He and Armstrong planted an American flag, gathered rocks, and set up experiments. After about two hours, they crawled back into the Eagle. Moon dust stuck to them, smelling like damp ashes. The men shivered with cold in the paper-thin LM and had trouble sleeping.

They spent 21 hours and 36 minutes on the lunar surface. To reduce launch weight, the backpacks and trash were dumped out the hatch. They also left the lower half of the lander. On the leg was a plaque that read, "We came in peace for all mankind."

As the Eagle rose up, the engine plume blew dust everywhere. "I looked up long enough to see the flag fall over," Aldrin recalled. After rendezvous in lunar orbit, Eagle was released to crash (location unknown) on the Moon. On July 24 the crew splashed down in the Pacific.

President Richard Nixon welcomed the crew on board the U.S.S. Hornet recovery ship. He said, "This is the greatest week

To reduce the mass (and thus the fuel required) for liftoff from the Moon, Apollo crews left white "jettison" bags (40" x 27") of trash on the surface like the one shown here. The bag could hold about the same amount as a lawn and leaf bag.

in the history of the world since creation." The astronauts listened from quarantine, where they were stuck for twenty-one days as a precaution against "moon germs." Mission Control filled with cigar smoke. On the big screen it said, "Mission Accomplished."

In Fun Is Where You Find It, Alan Bean recreated his toss of a piece of discarded foil insulation from the LM during Apollo 12. With no air on the Moon, it arced and plopped down instead of floating. The umbrella-shaped object is the radio antenna.

APOLLO 12 THE ARTIST'S MOON

The race was over, but both sides kept flying. In August, the Russians flew an unmanned Zond 7 around the Moon and returned without any failures.

This success paled in comparison with Apollo 12 that launched in November on a ten-day mission with Pete Conrad, Alan Bean, and Dick Gordon. After a precision landing only 535 feet from Surveyor 3, the men were able to walk over to it. They returned parts of it to Earth. Scientists were astonished to discover that bacteria from Earth had made the trip to the Moon, been exposed on its surface for about two and a half years and were still viable upon return to Earth. In 1991 Pete Conrad said, "I always thought the most significant thing that we ever found on the whole ... Moon was that little bacteria who came back and lived."

Conrad and Bean deployed the Apollo Lunar Surface Experiments Package (ALSEP) to study the environment of the lunar surface. The SEISMOMETERS on this and later flights revealed that the Moon's core is small: less than a quarter of its radius compared to Earth's core which is more than half its radius. Scientists also measured more than 1700 impacts, including spacecraft deliberately crashed to calibrate the seismometers.

Conrad with Surveyor 3

APOLLO 13 A SUCCESSFUL FAILURE

Apollo 13 launched in April 1970 with the crew of Jim Lovell, Fred Haise, and Jack Swigert. About half way to the Moon, an oxygen tank exploded in the service module of Odyssey. The astronauts shut down Odyssey to save its battery power for entry and used the LM Aquarius as a lifeboat.

But Aquarius was designed to support two men for two days, not three men for four days. They turned off the lights and most everything else to save power. Inside the Aquarius was like a damp cold cave. They had enough oxygen, but not enough disposable air filters to remove the carbon dioxide the men exhaled. Too much carbon dioxide is deadly. Odyssey had filters, but they were square instead of round. Flight controllers designed a contraption that looked sort of like a mailbox to mount these filters to the air system. Instructions were voiced up from Mission Control to make this device out of checklist pages, duct tape, plastic bags, and socks! Thankfully, it worked.

The crew swung around the Moon and made a heroic return to Earth. NASA called Apollo 13 "a successful failure" because the men survived, though it certainly wasn't the mission anyone had planned.

Apollo 13 Flight Controllers devised a "mailbox" contraption (held by Deke Slayton) using bags and duct tape to use square carbon dioxide filters to replace round ones in the lunar module.

APOLLO 14 PUSHING THE LIMITS

After the precision landing of Apollo 12, NASA decided to send Apollo 14 to more challenging terrain. On February 7, 1971, Alan Shepard and Ed Mitchell descended to a pock-marked area called Fra Mauro while Stu Roosa photographed the Moon from Kitty Hawk in orbit. Antares landed on the lumpy surface with one footpad in a crater, tilting the lander to one side. Thankfully, it didn't tip over.

But the landing almost didn't happen. Just 90 minutes before descent, a switch in Antares kept signaling to abort. When Mitchell tapped the panel, the signal would disappear. Mission Control guessed that a piece of SOLDER was floating around and causing an electrical short. If this happened during the descent, the computer would abort the landing. Could the computer be programmed to ignore the switch? Time was short. A young programmer from MIT named Don Eyles had to rewrite the code before Antares passed out of radio contact on the far side. He got it done in the nick of time. The landing was saved!

Shepard and Mitchell collected a cart load of rock samples their first day. The next day, they dragged their cart up the steep side of Cone Crater to get rocks thrown out from its impact. But with no landmarks on the unfamiliar terrain, they couldn't find the rim they sought. They pushed uphill in stiff suits until Mission Control forced them to stop. They learned post flight that they had passed just 65 feet from the rim.

One of the Apollo 14 ALSEP experiments found that rocket exhaust from landing created a temporary "atmosphere" for the airless Moon that lasted a few months. The experiment was sensitive enough to detect gas from the astronauts' space suits whenever they passed nearby.

Don Eyles seen here (right) with colleague Sam Drake, in the LM simulator at MIT in 1971, rewrote the program code that bypassed a faulty switch and allowed Apollo 14 to land on the Moon.

The Apollo 14 crew could not find the rim of Cone Crater though it was less than 100 feet away. The lack of trees or buildings for scale and the close horizon on the Moon make distance difficult to judge.

1970
April: Apollo 13
Sept: Luna 16 sample return*
Oct: Zond 8*
Nov: Luna 17/Lunokhod 1*
Dec: Kosmos 382*

1971
Jan: Apollo 14
July: Apollo 15
Sept: Luna 18*
Sept: Luna 19 orbiter*

1972
Feb: Luna 20 sample return*
April: Apollo 16
Dec: Apollo 17

1973
Jan: Luna 21/Lunokhod 2*
June: Explorer 49
pictured below

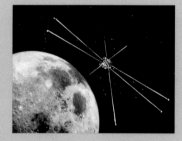

1974
June: Luna 22*
Oct: Luna 23*

1976
Aug: Luna 24 sample return*

*Russian missions

APOLLO 15 TRAINING PAYS OFF

Genesis rock

Despite a packed training schedule, Dave Scott and Jim Irwin took monthly geology field trips in the years prior to Apollo 15. They proved the value of this hands-on experience when they got to the Moon in July 1971.

Pebbles on the Moon begin life as high-speed objects thrown out by the blast of meteoroid strikes. They may end up hundreds of miles from their source. But a big rock is hard to move. A jolt strong enough to lift it is likely to break it into pieces. Thus, the larger the boulder, the more likely it is to have formed nearby. Knowing this, Scott was on the lookout for a boulder on the Moon. When he found one, he quickly identified it as a breccia, photographed it, and chipped off a sample. Then he rolled it over. The soil underneath, shielded from solar and cosmic radiation since the rock landed, revealed the age of the whole area.

Scott and Irwin were the first to drive a rover. They travelled miles from the shelter of their lander. On the way back from Mount Hadley, Irwin, born on St. Patrick's Day and of Irish descent, stopped for an unusual green rock. Soon after, the pair found the most important rock of all: a white rock sparkling in the Sun atop a pedestal. The men knew it was a chunk of the original lunar crust! Dubbed the Genesis Rock, it is 4 to 4.5 billion years old.

Al Worden spent dozens of hours flying over mountains and deserts in the western U.S. to hone his observation skills. He put those skills to use from lunar orbit. He noted layers in the white mountain rising from the dark lava of Tsiolkovsky Crater on the Moon's far side. He and other astronauts wished to one day visit this mysterious place. But with only two Apollo missions remaining, NASA considered it too dangerous.

Scott took this telephoto picture of Silver Spur showing distinct layers in the rock. This proved that the Moon had an extended period of volcanism.

During Apollo 15, Jim Irwin digs a trench to study the properties of the lunar regolith. Fine lunar dust clung to his spacesuit.

APOLLO 16 A HISTORY OF IMPACTS

With only two more missions, the geologists pushed to land at Tycho Crater, one of the youngest on the Moon. But a trip so far south of the equator required more fuel than any previous mission. An Apollo 13 type of rescue would be impossible. NASA said no. So the science team chose the lunar highlands.

The highlands were thought to have formed from a different, thicker kind of lava containing more sand than the maria. Mountains of this lava would harden into a stone-like granite and be a lighter color. They would form elongated ranges with irregular shapes and rounded tops: like those seen near Descartes Crater, west of the Sea of Nectar. The science team sent Apollo 16 there to collect volcanic rocks.

John Young, Charlie Duke, and Ken Mattingly flew to the Moon in April 1972. Mattingly observed from Casper in orbit while Young and Duke took Orion to the surface. Over three days, they collected 731 samples, including a deep drill core from 7.2 feet below the surface.

But none of the rocks were volcanic. They were breccias, refugees that had fled violent impacts from hundreds to thousands of miles away. Some came from Theophilus Crater in the Sea of Nectar about 155 miles from the landing site. These were rich in titanium and aluminum, dating the lava in the Sea of Nectar to 3.79 billion years, and the impact that created the Sea at 3.92 billion years. Even though the Sea of Rains is 620 miles away, it was such a huge impact that some EJECTA landed at Descartes 3.8 billion years ago.

The history of the Moon turned out not to be one of volcanism, but instead a series of violent impacts.

Even the boulders were not from around here: This breccia brought back by Apollo 16 was a piece of a boulder four feet high and five feet long. The fragment is very hard with a variety of inclusions including bits of metal.

The Apollo 16 Lunar Surface Experiment Package (ALSEP) included L to R: a nuclear battery, the central station (gold), and the passive ("deflated balloon" is a sun shade) and active (orange flag marker) seismic (moonquake) experiments.

APOLLO 17 A FOUNTAIN OF KNOWLEDGE

During Apollo 17, Cernan accidentally caught a hammer on the right rear fender, breaking off part of it. He taped it back on, but it fell off while driving. Lunar dust got all over everything. Mission Control helped the crew make a fender out of maps, clamps, and duct tape. It held through 18 miles of driving. Cernan brought this makeshift fender back to Earth. It is on display at the National Air and Space Museum.

In 1964, Geologist Eugene Shoemaker created the first geologic map of the Moon and established the U.S. Geological Survey's first office of astrogeology in 1965. Harrison "Jack" Schmitt was working for him when he was selected as an astronaut. He would be the first scientist to walk on the Moon.

Cernan and Schmitt landed the Falcon in the Taurus-Littrow Valley on the eastern edge of the Sea of Serenity in December 1972. While Ron Evans orbited in America, the last two Moon walkers drove about 21 miles in their rover and collected 245 pounds of samples.

These samples revealed unexpected details of the Moon's history. Near Shorty Crater, Schmitt took a photo of an area showing a boulder he and Cernan would

soon sample. Schmitt looked down at his feet. His boots had scuffed away a layer of gray to reveal colored soil underneath. "It's orange!" Schmitt said. He first thought the soil had been oxidized, rusted, like orange soil on Mars. But lab tests showed it formed in a much more exciting way: via a fire fountain.

As lava came to the surface, gases bubbled out like shaken soda pop, propelling the molten rock hundreds to thousands of feet into the lunar sky. During flight, they cooled into orange, red, and black glass beads that rained down slowly in the low gravity. Those rich in iron were black and near the bottom of the core sample, accounting for why the maria are so dark. The orange beads were 3.64 billion years old and formed from material about 250 miles underground. They'd been buried under later lava flows, then exposed 19 million years ago when an impact created Shorty Crater.

Another surprise came all the way from Tycho. Even though it was 1300 miles to the south, a ray stretched across the Apollo 17 site. Geologists speculated that debris from the impact had triggered a landslide here. A sample scooped up by the crew indicated that the presumed Tycho event was about 109 million years ago.

Apollo 17 brought an end to the first human explorations of the Moon. We had learned a lot, but the Moon still had much to teach us.

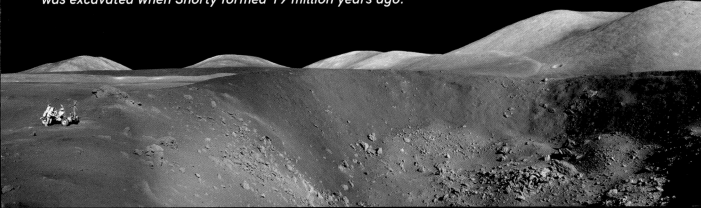

On the edge of Shorty Crater Harrison Schmitt discovered the remains of a fountain of lava that erupted 3.5 billion years ago and formed orange soil, in front of the large boulder, that was excavated when Shorty formed 19 million years ago.

The Russian Luna 17 lander is still sitting on the Sea of Rains where it delivered the world's first robotic lunar rover, Lunokhod 1, in November 1970. A closeup shows the tracks of the rover circling the lander.

SOVIET LUNAR EXPLORATION

After the success of Apollo 11, the Soviets pretended that they never planned to send men to the Moon—only robots. Luna 16 was the first robot to return samples from another world. It brought soil back from the Sea of Fertility in September 1970. Luna 17 deployed the first lunar robotic rover, Lunokhod 1, in November. This hardy robot roamed the Sea or Rains for eleven months and took over 20,000 photos.

The Russian unmanned Zond 8 looped around the Moon and returned to Earth in October 1970. A December test of their lunar lander on Kosmos 382 proved its design worked. But the Russians couldn't afford to compete with the Apollo program. They cancelled their L1 lunar module without it ever having flown a cosmonaut.

The Russians focused their human program in LEO (building the first space station in 1971) while continuing robotic studies of Venus, Mars and the Moon. Luna 18 crashed during landing in the Sea of Fertility in September of 1971. Luna 19 studied the Moon from orbit starting that October. In February 1972, Luna 20 landed on the Moon, retrieved a soil sample from the Sea of Fertility, and returned to Earth. But this feat was hardly noticed as Apollo 16 roared to the Moon in April.

In January 1973, the Russians landed Luna 21 in LeMonnier Crater not far from where Apollo 17 had been just a month earlier. The Lunokhod 2 rover hibernated during the lunar nightspan, heated by a radioactive power source. With eight wheels

HISTORY OF EXPLORATION

for traction, it crawled 23 miles across the surface, surveying the chemistry of the surface and measuring solar x-rays. Unfortunately, in its fifth month, it rolled into a crater. Dust settled on its solar panels and radiators. Before it failed, it had transmitted more than 80,000 television pictures. Its laser retroreflector is still used by Earth-bound lasers to measure the distance to the Moon.

Luna 22 entered orbit in June 1974 and operated for 18 months. Luna 23 landed in October. But it fell on its side and couldn't collect a sample.

Luna 24 was the last mission by any nation to land on the Moon in the 20th century. After landing in view of its predecessor (Luna 23) in the Sea of Crises in August 1976, it drilled 88" into the soil. The soil was stored in a flexible tube that was coiled into a spiral on a drum and sealed inside a metal can. About 6 ounces were returned. A tenth of an ounce was shared with NASA scientists in Houston. The sample was found to be high in aluminum and low in titanium.

The U.S. and Russian programs pushed the miniaturization of electronics, high-strength, heat resistant, and low-weight materials and rocket engines. These spinoffs of the space race led to new technologies such as medical instruments, supermarket barcodes, and home computers. The knowledge gained about the Moon led to a wealth of new discoveries and theories about the solar system that would continue to be refined in the coming decades.

The Luna 24 spacecraft landed in the Sea of Crises in August 1976. Using a sample arm and drill, it collected a sample that was determined to be only about 300 million years old.

HISTORY TOLD BY APOLLO ROCKS

Rocks brought back from the Sea of Tranquility were basalts: proving the maria are seas of dark lava. Radioactive ISOTOPES showed this lava flowed 3.65 billion years ago. Apollo 12's basalts cooled into rock 500 million years after that. (They contain less titanium which accounts for the more reddish color of that region compared to the Sea of Tranquility.) These different ages implied that the Moon had a liquid mantle between the crust and core for an extended period of time.

Apollo 14's samples were mostly breccias, rocks formed via impacts. The pieces of basalt in these rocks were 4.0 to 4.3 billion years old, pushing back the age of the Moon. The oldest was recently dated via ZIRCON crystals at 4.51 billion years, older than Apollo 15's Genesis Rock. Apollo 15's rocks revealed that the Sea of Rains impact occurred between 3.84 and 3.87 billion years ago.

The landing area of Apollo 16 was thought to be a volcanic plain, but it turned out to be debris from impacts. The rocks dated the Sea of Nectar's impact at 3.92 billion years ago, and the lava that filled it at 3.79 billion years. One Apollo 16 breccia contained the oldest piece of original lunar crust dated at 4.51 billion years old.

The Sea of Serenity formed later than the Sea of Nectar but prior to the Sea of Rains around 3.8 billion years ago. Seismic and gravity data indicated the basalts found at the Apollo 17 site represent a layer of lava that is 0.6 to 0.8 miles thick. Apollo 16 and 17 samples had bits of rock ejected from Tycho, dating its formation to about 108 million years ago.

The Apollo rocks showed that the Moon had a violent past and the large basins formed between 3.9 to 3.8 billion years ago. This period was named the Late Heavy Bombardment or Lunar Cataclysm. This period of solar system evolution may have been key to delivering water, and hence life, to Earth.

John Young examines a giant breccia during Apollo 16.

CONNECT THE APOLLO DOTS

Here's how to find the six Apollo landing sites:

1. The equator: 11, 12, 14. Look for the upside-down rabbit shape with the "head" being the Sea of Tranquility and with the Seas of Fertility and Nectar as the "ears." Apollo 11 is where the Sea of Nectar "ear" attaches. Draw a dotted line west from there along the lunar equator. Just below this line and "under" the bright dot of Copernicus Crater are Apollo 12 and 14.

2. The "N:" 15, 16, 17. From 14, draw a line up to between the "eyes" of the Seas of Rains and Serenity. The "bridge" of the Moon's "nose" is Apollo 15's site. Connect this "dot" to Apollo 16, below the equator, just west of the bend in the Sea of Nectar "ear." Finally, complete the (somewhat slanted) letter "N" by drawing a line from 16 to Apollo 17 on the upper notch between the Sea of Tranquility (rabbit's head) and the Sea of Serenity. See the Animated Moon Map at http://www.MarianneDyson.com/moon.html.

Next full Moon, see who else "nose" how to find the six Apollo sites!

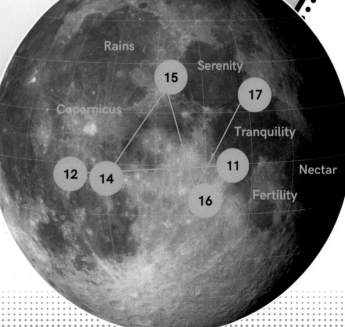

CHAPTER 3
NEW VIEWS OF THE MOON

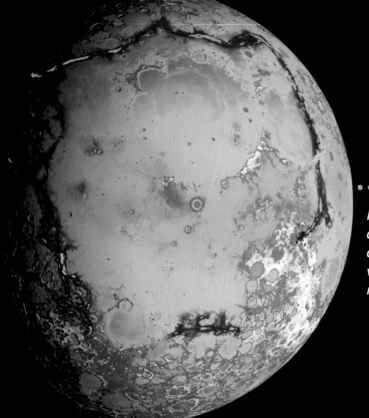

NASA's GRAIL spacecraft revealed masses hidden beneath the surface (low elevation in blue and high in yellow) of the Moon such as lava-flooded rift valleys (red-brown) that form a rectangular border around the Ocean of Storms.

Apollo samples allowed scientists to plot a timeline of lunar history. A Giant Impact created the Moon (with a small iron core) out of the Earth's mantle about 4.5 billion years ago. The molten Moon (lacking water) formed a crust of anorthosite. As tides slowed the Moon's spin, the near side faced Earth permanently. The large lunar basins formed during the Late Heavy Bombardment 3.9 to 3.8 billion years ago. As the Moon cooled and shrank, mantle material erupted where the crust was thin under the basins. For billions of years, layer upon layer of lava filled these low places and formed the "lunar bedrock" seen by Apollo 15 and 17 astronauts. Violent volcanic fire fountains explained the dark maria as well as green and orange glasses found in samples. Impacts (including Tycho 100 million years ago) sprayed dust everywhere and coated the highlands with a thick blanket of regolith. The Sun and COSMIC RAYS darkened the surface over eons. Yet the Moon still held secrets. Like failing to see a hidden picture until told to look for it, scientists saw Apollo's glass beads, but didn't appreciate their significance for decades. Improved technology and international cooperation led to better maps and more ways to probe the mysteries of the Moon. Starting in the 1990s, scientists would learn that the Moon's treasures had barely been sampled by Apollo.

Clementine photographed the Earth rising over the north pole (Plaskett Crater) of the Moon. Clementine orbited the Moon for two months in 1994 and provided the first hint that the Moon contained hidden resources at the poles.

HIDDEN IN SHADOW

The first hints of the Moon's hidden treasure came in 1996. The U.S. military had sent Clementine to the Moon to test new spacecraft technology in 1994. The spacecraft successfully mapped nearly 100 percent of the lunar surface using 11 different filters (see Mapping the Minerals). One particular pass over the south pole provided a big surprise. In 1996, the team reported that radio echoes from Shackleton Crater suggested the presence of water ice. (Ice reflects radio waves differently than soil.) How could this be? Ice would SUBLIMATE away on the airless Moon. The gravity was too weak to hold it down.

But wait—because the Moon's equator is within 5° of the Sun's equator (unlike Earth's which is tilted 23.5°), the Sun never passes over the poles. As the Moon spins, the Sun remains on the horizon and appears

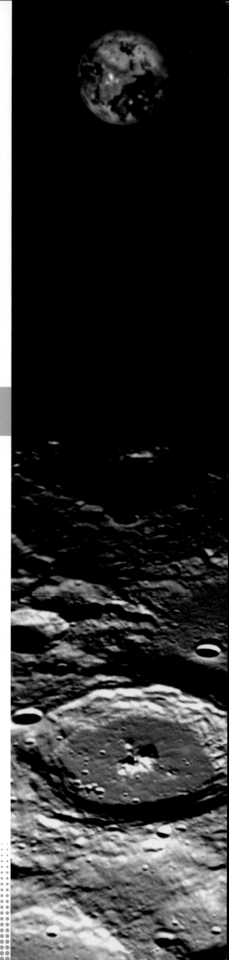

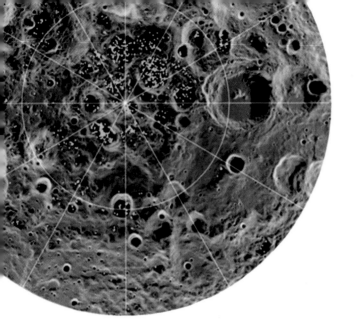

Clementine provided the first evidence of water at the south pole. This was confirmed (blue dots) by NASA's Mineralogy Mapper in 2018.

to move in a circle around the poles. Deep polar craters never see the Sun or feel its warming rays. The temperature in these craters is near absolute zero at -387°F. If water sprayed out from the impact of a comet and plunged into one of these "wells"—it would freeze instantly. So ice was *possible*. But proof required more data.

In 1998, Lunar Prospector provided more evidence. It found hydrogen at both poles. There appeared to be about 6.6 billion tons of ice buried under about 18" of dry regolith.

But the hydrogen maps were not of sufficient resolution to tell if the hydrogen was concentrated as ice in dark craters or spread out in the regolith (implanted by the SOLAR WIND). Lunar Prospector was purposefully crashed in Shoemaker Crater near the south pole in the hope water could be detected in the plume. However, none was observed.

With NASA's attention on Mars and completion of the International Space Station, international cooperation would be required to finally answer the water question.

Lunar Prospector carried an aluminum capsule with the ashes of Geologist Eugene Shoemaker who trained Apollo astronauts and who was the first, and so far only, person to be buried on the Moon.

MAPPING THE MINERALS

Light comes in a spectrum of colors from red to purple that represent bands of different wavelengths of light. Red has the longest wavelength, and violet (purple) has the shortest. White objects reflect all wavelengths, and black objects absorb all of them—which is why a white car in the Sun is cooler than a black car next to it.

The Clementine spacecraft cameras used filters to block all but one band of color at a time—snapping multiple photos in each of 11 different bands. Scientists used knowledge of which colors minerals absorb to determine the composition and history of the surface of the Moon.

Lunar scientist Paul Spudis applied this technique to the Moscow Sea on the lunar far side. He started with a black and white image (1) that shows the brightness or ALBEDO of the surface. The white highlands reflected the most light, and the black maria absorbed the most. So these two types of terrain are easily distinguished by albedo.

When different color bands are overlaid on the albedo map (2), the black maria divides into two distinct areas. The west/left side is high in iron (green). The right/east side is high in titanium (blue) which absorbs light at a shorter wavelength. The highlands are low in titanium and high in volcanic glass (red). Fresh craters also show up as blue.

Plotting the abundance of titanium (3) adds age information. The low titanium lava (blue and cyan) erupted first from the edge of the basin and flowed downhill toward the center. Then high titanium (yellow and red) lava came later from vents on the eastern side at higher elevation.

Every time scientists point a new type of camera at the Moon, more details about its composition and history are revealed.

Moscow Sea

THE INTERNATIONAL MOON

Japan's Kaguya (Moon princess) sent this final image just before impact in 2009.

Japan had become the third nation to reach the Moon in 1990. Hiten (Celestial Maiden) was designed to gain experience with spacecraft. After completing all its tests, Hiten was purposefully crashed on the near side south of the Sea of Nectar in April 1993. A decade later, Japan developed an ambitious lunar mission (SELENE, renamed Kaguya). The size of a school bus, Kaguya launched in 2007. It provided a high-quality elevation map. However, the data was not shared with the international community until late 2009, lessening its impact.

The European Space Agency (ESA) flew a small technology demonstration, SMART-1, in 2003. Testing a new solar electric propulsion design, it took a year to reach the Moon. Its science instruments were limited, but it mapped the poles during a different season than Clementine. SMART-1 thus confirmed the poles have permanently-shadowed craters and peaks that are almost continuously sunlit.

In October 2007, China became the fourth nation to reach the Moon. Chang'E-1 (named for a mythic Moon goddess) generated a high definition 3D map. A second orbiter in October 2010 added detail to their maps and also did tests for future missions.

But RADAR, which can peer into dark craters, was needed to answer the water question. So an American team developed a radar experiment for the Lunar Reconnaissance Orbiter (LRO). They also flew a small version, called Mini-SAR, on India's first lunar spacecraft, Chandrayaan-1 (lunar vehicle) which launched in 2008.

The Chandrayaan radar experiment convinced all but a few skeptics that water ice existed in polar craters. LRO would provide the final confirmation.

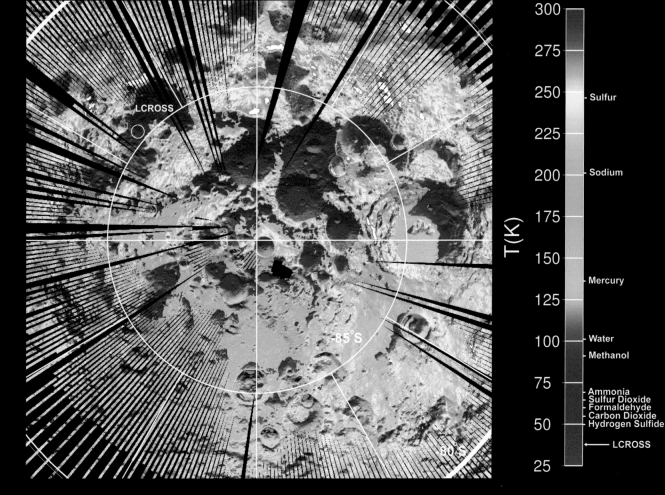

LRO's map of the lunar south pole shows that even near their annual maximum temperature, some craters are so cold that they can keep compounds such as water ice and methane (possibly delivered by comets) frozen for more than a billion years. (25 to 300°K = -248 to 27°C and -414 to 80°F.) The LCROSS spacecraft was purposefully crashed into one of the coldest of these craters.

Launched in June 2009, LRO generated a thermal map of the south polar area. In October, the LRO team orchestrated a spectacular crash. First, an empty rocket casing was aimed at Cabeus Crater near the south pole. Its impact sent up a spray of ejecta from the crater's floor. Then, the Lunar Crater Observation and Sensing Satellite (LCROSS) followed four minutes later, measuring and passing through the EJECTA. LCROSS data revealed a concentration of about 6 percent water in the impact area. Finally, ice was confirmed! But the Moon had more secrets to share.

WELCOME TO THE MOON

In 2008, a new look at Apollo rocks revealed water trapped in volcanic glass beads. Then Chandrayaan-1 revealed that water ice also exists in low concentration over much of the lunar surface, not just in deep craters. Where did this water come from? How deep does it go? Missions in the 2010s were primed to answer these questions.

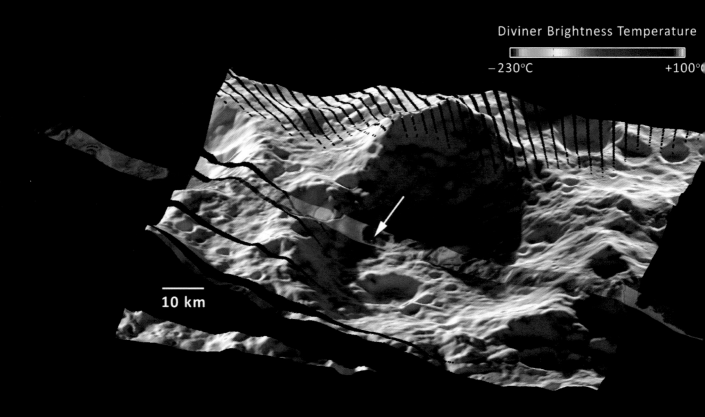

This image of Cabeus Crater shows the temperature of the terrain where LCROSS impacted. About 90 seconds after the impact, the site (white arrow) was heated to more than 1300°F (700°C).

DEEP SECRETS

A spacecraft placed in lunar orbit is constantly tugged on by the Moon's gravity. During Apollo, scientists found that some places on the Moon tug harder than others. Areas where mass is concentrated have stronger gravity such as where impacts have compressed the ground. These mass concentrations, or MASCONS, pull spacecraft forward, back, left, right, and down—making most low lunar orbits unstable. For example, mascons caused a small satellite released by Apollo 16 to crash after only 35 days.

NASA launched the Gravity Recovery and Interior Laboratory (GRAIL) mission in 2011 to map the location and strength of mascons. GRAIL consisted of twin satellites named Ebb and Flow that were placed in a low lunar orbit of only 34 miles. The distance between the two satellites varied

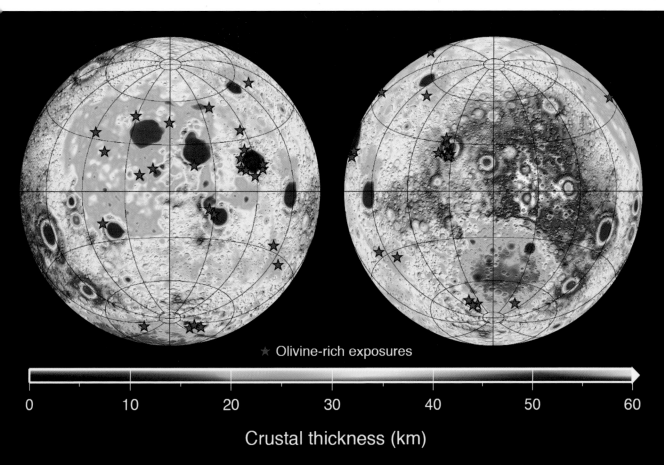

★ Olivine-rich exposures

Crustal thickness (km)

This map based on GRAIL data shows that the Moon's crust is up to 37 miles (60 km) thick (white) on the far side. Impacts compressed and thinned the crust under large impact craters (blue and purple). The Sea of Rains, Tranquility, and Crises form a row of low spots across the northern near side (left image). The crust is so thin at the Sea of Crises and the Moscow Sea (upper far side-right image) that olivine (purple stars), a mineral from the Moon's mantle mapped by Kaguya is exposed on the surface

WELCOME TO THE MOON

slightly as they flew over areas of greater or lesser gravity caused by masses hidden under mountains and craters.

Large mascons were found underneath all the biggest impact craters on both the near and far sides. (None were expected, and none were found under the Sea of Tranquility which doesn't have a ring around it like the Seas of Rains and Serenity.)

But a surprise lurked under the Ocean of Storms. Instead of a mascon, the gravity map revealed long thin features forming a rectangular "box" around the area. Scientists think these are ancient RIFT VALLEYS where the crust spread apart. Lava poured out of these rifts and flowed into the low-lying area between them, creating the Ocean of Storms. How did these rift valleys form? Some scientists speculate they are fractures caused by the impact that created the South Pole-Atkin Basin on the far side. GRAIL's data also proved that the crust of the Moon is indeed thicker on the far side. The reason for this difference remains a mystery for future lunar pioneers to solve!

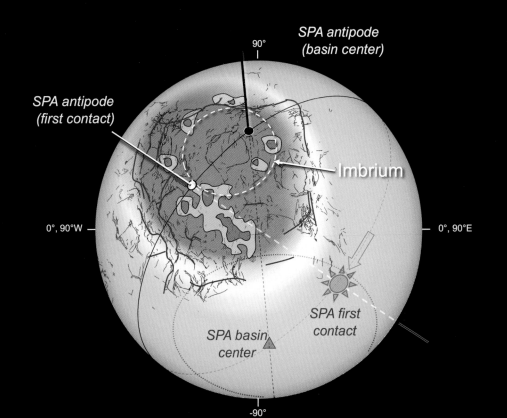

The impact that created the South Pole Aitkin (SPA) basin is on the opposite side (antipode) of the Moon from the Ocean of Storms (orange) and Sea of Rains (Imbrium-outlined in white). Could the impact have caused the rectangular deep intrusions (blue outline) discovered by GRAIL?

In 1968, Surveyor 7 captured the horizon's glow just before dawn in Tycho Crater.

GLOWING NEON SIGNS

In the 1960s, NASA cameras on robotic landers captured a bright glow during lunar sunsets. Apollo astronauts orbiting the Moon also reported seeing a glow above the surface. This glow comes from the Moon's ultra-thin atmosphere, called an EXOSPHERE. In 2013, NASA launched the Lunar Atmosphere and Dust Explorer (LADEE, pronounced "laddie") to learn what the exosphere is made of.

Scientists guessed that the exosphere was mostly created by solar wind gases that smack into the Moon at about a million miles per hour. The heavier particles "stick" to the surface, and the lightest ones: helium, neon, and argon, "bounce" off into the exosphere. LADEE confirmed this hypothesis.

But the solar wind isn't the only source of gases. LADEE also found gases released from lunar rocks. Some increased and then decreased 25 percent over the 6-month mission, possibly as the result of tidal stresses on the Moon.

And then there is the dust! About 260 pounds of dust gets kicked up every day. This dust forms a permanent, but lopsided, cloud hugging the Moon. The cloud hangs over the TERMINATOR—the line between day and night.

Scientists think that most of the dust comes from the remains of old comets passing through the solar system—the source of annual meteor showers on Earth. Comets fall toward the Sun from a much farther distance than asteroids, and thus are moving much faster. Just one high-speed particle can scuff up a cloud of thousands of dust grains. The dust arcs up to 62 miles and falls back to the surface in about 10 minutes.

BACK TO THE SURFACE

Though most of the modern lunar spacecraft were purposefully crashed on the surface of the Moon (Hiten, Lunar Prospector, Kaguya, LCROSS, GRAIL, and LADEE), it wasn't until 2013 that a modern spacecraft actually landed on the Moon.

In December 2013, the Chinese Chang'E-3 (mythical Moon Goddess) carrying the Yutu (Jade Rabbit) rover landed in the Bay of Rainbows. This sophisticated rover transmitted video in real time while digging and performing analysis of soil samples. By hibernating through 32 lunar nightspans using plutonium heaters, Yutu was able to explore 1.2 square miles.

Chang'E-3 provided "ground truth" to measurements obtained from orbit. For example, the temperatures measured on the surface were higher than predicted by the models based on orbital data. They also found evidence that the Sea of Rains basin is subsiding, possibly because the

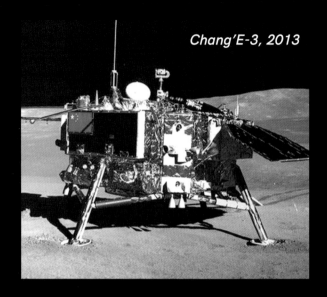

Chang'E-3, 2013

Moon is still cooling and shrinking. The scientists were excited to find some rocks that appear to be a new kind of basalt, different in composition from the Apollo and Luna samples.

After the success of Chang'E-3, the Chinese launched an even more ambitious mission: Chang'E-4, the first spacecraft to

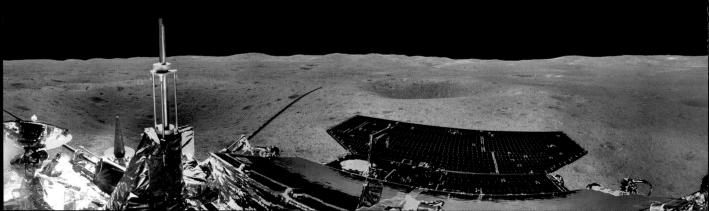

land on the lunar far side. Communications with the far side require a relay satellite which the Chinese launched in May 2018. The lander launched in December and sent back the first photos from the surface of the far side on January 2, 2019. Von Kármán crater was selected because it is relatively flat, a safe place for a first far side landing. It is also within the SPA basin which is perhaps the oldest surviving impact crater in the solar system.

Chang'E-4 may still be collecting data when its larger cousin launches in late 2019. Chang'E-5 is planned to drill and return a soil sample from the Mons Rümker region on the near side.

Israel launched its first lunar spacecraft, Beresheet (Genesis in Hebrew), in February 2019. Beresheet was built using private funds and gifted to Israel to inspire Jewish youth. The 1300-pound spacecraft spiraled its orbit outward by swinging around the Earth. It successfully entered lunar orbit, but it crashed in April attempting to land.

Chang'E-4, 2019

Spacecraft from multiple nations have now uncovered many previously hidden lunar resources—especially water ice that is concentrated near the poles but available in small amounts everywhere on the surface. Many questions remain. But one thing is certain: the Moon has the resources necessary to support human space exploration, and potentially, settlements. The biggest question is how soon can we get there?!

CHAPTER 4
HUMAN RETURN

Now that the resources of the Moon are better understood, and spacecraft are again actively exploring its surface, multiple nations are making plans for human missions. Political, financial, and technical challenges continue to alter the details and timing of these plans. But the basic requirements to fly to the Moon and back have not changed since the Apollo era.

The Chinese Long March 5 is one of the world's most powerful boosters, shown in 2017 at the Wenchang Space Launch Center on Hainan Island. The characters on the rocket mean Chinese space flight.

STAGING THE LAUNCH

Escaping from Earth requires massive amounts of fuel. About 90 percent of the weight of the Apollo Saturn V was fuel.

Modern rockets are also mostly big cylinders of fuel with engines to burn it. As the burned fuel shoots out the tail, the rocket is thrust upward. The Saturn V had three stages stacked on top of each other, each with fuel and oxidizer tanks and engines at the base. Once the fuel was used up in each stage, the stage and its engines detached (via bolts that exploded) and fell into the ocean. Because dropping stages reduces the remaining mass to be lifted to space, modern rockets still use stages. Rockets launch over oceans or uninhabited areas to prevent spent stages from falling on people.

Many rockets also use "strap on" boosters. These boosters are mounted around the "core" booster rather than making the "stack" taller. SpaceX boosters fly themselves back to a platform so they can be reused. Reusing boosters saves manufacturing costs and avoids contaminating the ocean with rocket parts.

The type and number of engines varies, but all rockets use similar fuels (see Gassing Up, page 67). The Saturn V had five engines that used LOX to burn kerosene in its first stage. The second stage had five smaller engines and burned liquid hydrogen (LH2). The third stage had just one engine that burned LH2. The Apollo rockets were the first to use very cold, CRYOGENIC, liquid fuels.

TO THE MOON AND BACK APOLLO STYLE

Everything (people, fuel, food, water, air, batteries, experiments, rovers, tools) needed for each mission was launched on one Saturn V rocket. During the Ascent (1), the Saturn V dropped two of its stages as their fuel was used up. The third stage, with the LM inside and the CSM on top then entered low Earth orbit (2). The third stage was fired to reach escape velocity and then discarded. (3) The CSM engine was used to turn around and dock with the LM, discarding its launch shroud.

The two vehicles flew nose to top while rotating in "barbecue" mode during Trans-Lunar Coast (4) lasting about two days. Power was provided by batteries called FUEL CELLS. The CSM had no bathroom—the men vented urine overboard and stored solid waste in bags.

At the end of Trans-Lunar Coast, the

APOLLO TRAJECTORY PROFILE

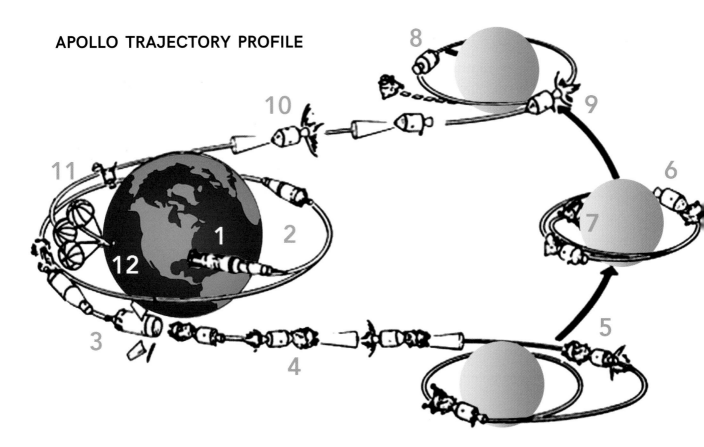

The five engines of the Saturn V first stage burned 203,000 pounds of RP-1 and 331,000 pounds of LOX in 2.5 minutes to generate a thrust of a million pounds. The Apollo 11 first stage separation (shown) was at an altitude of about 38 miles. The stage fell into the Atlantic Ocean about 55 miles from Kennedy.

service module engine was used to slow down into Lunar orbit (5). The LM with two crewmembers separated from the CSM and used the Lunar Descent Module engine for the lunar landing (6). The LM had no airlock or bathroom. The whole module had to be evacuated of air for each extravehicular activity (EVA) during lunar surface operations (7). The LM and spacesuits and rovers used non-renewable battery power. The LM descent module, EVA backpacks, rovers, experiments, tools, trash, and mementos were left on the surface.

The LM ascent engine provided power for the lunar ascent and rendezvous (8). After docking, the men transferred themselves and the lunar samples to the CSM. Then the LM was discarded to crash on the lunar surface. The CSM engine was used to escape lunar orbit (9) and enter the Trans-Earth Coast (10).

The service module was discarded prior to reaching Earth's atmosphere. Only the command module capsule entered the atmosphere. The heat shield on the CM ABLATED, or burned off, during the first part of the Earth entry (11). Two sets of small parachutes were deployed. These were fol-

The command module (Apollo 11 shown here on the deck of the U.S.S. Hornet) is the only part of the original launch stack that returned to Earth after each Apollo flight. Note the air bags, flotation collar, and blackened heat shield. This capsule is on display at the National Air and Space Museum.

lowed by three main (red and white) parachutes to slow the module before it splashed down in the ocean (12). At splashdown, balloon-like balls were inflated to keep the capsule upright. Navy "frogmen" then attached flotation collars to the capsules. The capsules and crew were lifted by helicopters onto the deck of aircraft carriers during the recovery phase.

MOVING OUT OF EARTH ORBIT

Unfortunately, no current rocket can match the Saturn V's lift capability of 310,000 pounds to low Earth orbit. It will be many years before new rockets such as the Space Launch System (SLS), Elon Musk's (SpaceX) Starship, and Long March 9 offer a similar capability.

But giant rockets are not needed to reach the Moon! Instead, the PAYLOAD (supplies and vehicles) and crew can be launched separately and rendezvous in orbit. Space station partner nations have decades of experience doing business this way. Cargo and crew ride up separately to the ISS. Crews ride Russian Soyuz TMA "taxis" while supplies go up on Russian, American, European, and Japanese cargo rockets. Boeing's CST-100 Starliner and SpaceX's Dragon 2 capsule are expected to launch crew within the next year. Blue Origin, run by Amazon's Jeff Bezos, is also developing a crew-capable rocket called New Glenn. It's expected to fly in 2021.

China is also separating their crew and cargo launches. A variant of their Long March 5B is expected to launch new space

Instead of one Saturn V-sized rocket, multiple launches of the Chinese Long March 5 (see page 60), the American Delta IV Heavy (left), SpaceX Falcon Heavy (center), and Russian Proton M (right) could support lunar missions.

station modules in 2020. A similar rocket will launch a Shenzhou capsule with crew separately.

Lunar cargo includes modules, supplies, and fuel. The mass of the Apollo command, service, and lunar modules at launch was about 67,000 pounds. Studies show that by using modern materials, this weight could be reduced by thousands of pounds. Launching without fuel (and gassing up at a space station) would reduce the weight by about half.

Multiple heavy-lift rockets are available now to boost lunar cargo. The American Delta IV Heavy can lift about 63,000 pounds, as can SpaceX's Falcon Heavy. The Chinese Long March 5 and the Russian Proton M and the SpaceX Falcon 9 can lift more than 50,000 pounds to LEO. If needed, smaller rockets such as the Atlas V, Antares, European Ariane 5, Japanese H2, and Indian Polar Satellite Launch Vehicle could loft extra fuel or supplies.

So lunar missions do not have to wait on new rockets. But they do have to wait on new crew modules. No spacecraft can currently support people for long periods in deep space or land them on the Moon and bring them back to Earth.

The SLS will lift Orion to the Moon using the world's largest solid rockets (one shown during a 2016 test that blasted 3.6 million pounds of thrust in two minutes) built by Northrop Grumman Innovation Systems.

GASSING UP

Hydrogen is the lightest element and burns at the highest temperature of any known substance. So hydrogen generates more energy for its mass than any other propellant. However, storing it is a challenge. LH2 must be kept at -423°F. Heat causes it to change into a gas and expand, and potentially explode its container. Also, because it is the smallest of molecules, it can leak through even the best of welds or seams in holding tanks.

Rocket Propellant-1 (RP-1) or refined kerosene, is derived from petroleum and is stable at room temperature. Tanks of RP-1 don't have to be insulated and aren't likely to leak. RP-1 doesn't produce as much energy as hydrogen by weight, but it produces about four times as much energy by volume because it is more dense than LH2. One gallon of RP-1 weighs as much as 11 gallons of LH2.

Some boosters, like those used by the Space Shuttle and planned for the SLS, use solid fuels where the oxidizer and fuel are mixed together. These fuels are easier to store than liquid fuels. But once solid rockets are "lit" they can't be throttled or shut off, limiting abort and landing options.

The Saturn V, Russian rockets, and the Chinese Long March use RP-1 for first stages where pushing power is more important than weight. Above the atmosphere, they use LH2 because it is light weight and efficient. Both fuels use LOX to burn them. LOX and LH2 can be manufactured out of water on the Moon in the future, allowing rockets to be refueled, and reused in space.

Falcon 9 has a reusable first stage with nine Merlin engines that burn RP-1 to boost the payload. After separation, it flies back to land on a drone ship platform in the Atlantic Ocean. The Falcon Heavy has three of these boosters.

NEW CREW MODULES

Author Marianne Dyson poses with an Orion test vehicle at the Pima Air & Space Museum in Arizona in 2013. The full-scale mockup was on its way to California for splashdown tests. A similar mockup flew into space atop a Delta IV Heavy in 2014.

The American Orion, Russian Federatsiya (Federation), and private company SpaceX's Dragon 2 are all designed to carry crew beyond Earth orbit. The SLS 1B, Soyuz 5, and SpaceX Starship boosters to lift them are being developed in parallel.

At 16'6" in diameter, Orion is larger and heavier than an Apollo CM (12'10" in diameter). It can support four to six crewmembers for up to 21 days versus Apollo's three astronauts for up to two weeks. In place of Apollo's disposable filters and batteries, Orion has a reusable carbon dioxide removal system and solar arrays. Computers can "fly" Orion, so no astronaut must stay in space while others land on the Moon. And unlike Apollo, Orion will have a toilet! The first human flight of Orion, Exploration Mission-2 (EM-2) is planned for 2022. Later missions may take crew to a proposed Gateway space station in high lunar orbit. (Orion can't land on the Moon.)

The Russian Federation is 14'8" in diameter and designed for four to six people. Federation will also have solar arrays and support missions of about ten days. Unlike the current Russian Soyuz TMA, Federation's docking port will be reused for 10 flights. Also, it won't require the costly custom-fitted seats of the Soyuz. Human launches are expected in the mid-2020s, depending on heavy-lift booster availability.

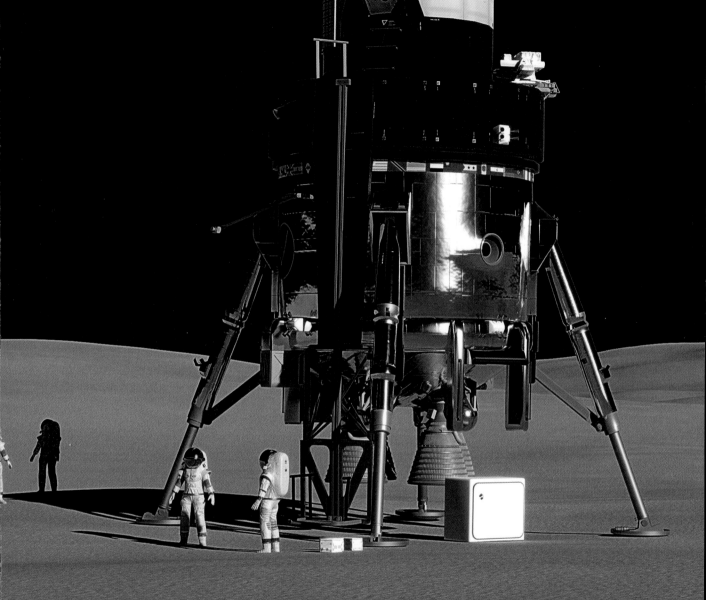

In 2018, Lockheed Martin proposed a lunar module. It would take a crew of four to and from NASA's proposed Gateway and support them for 2 weeks. An SLS booster would fly it to Earth orbit. The module would then fly to the Gateway to be refueled with LH2 and LOX, possibly from the Moon.

Federation would support human CIRCUMLUNAR missions in the 2025 to 2030 timeframe.

In 2018, SpaceX announced plans to send Japanese billionaire Yusaku Maezawa around the Moon as soon as 2023. The Dragon 2 crew module can carry up to seven astronauts to LEO. Its heat shield is already designed to handle lunar return velocities. At 12 feet in diameter, it is about the size of an Apollo capsule. It is light enough for the Falcon 9 booster to take it to LEO. To go to the Moon requires a more powerful third stage such as in a Falcon Heavy or Starship.

Reaching the surface of the Moon, staying there, and launching again requires new lunar modules. NASA plans to start flying small commercially-provided robotic landers to the Moon in 2021. NASA's goal is to have humans on the surface between 2024 and 2028.

Russia plans to restart their robotic lander series with Luna 25 in the next few years. A sample return, Luna 28, would launch 3 to 7 years later. Luna 28 might rendezvous in lunar orbit with a Federation crew launched separately. Crew may then remotely control rovers on the Moon and choose samples to bring back to Earth.

In 2017, Japan announced plans to put an astronaut on the Moon around 2030. Japanese car company Toyota is now working on a human lunar rover.

The European Space Agency revealed plans to mine the Moon's regolith by 2025. They are working with industry on using 3D printing to build a lunar base.

China already has multiple rovers on the Moon. They are developing a next generation crew capsule to support their space station and lunar missions. The crew would launch on the Long March 5B and rendezvous in LEO with the lunar module launched on a Long March 9. Their long-term plans include a research base at the lunar south pole by 2030 that will support human visits.

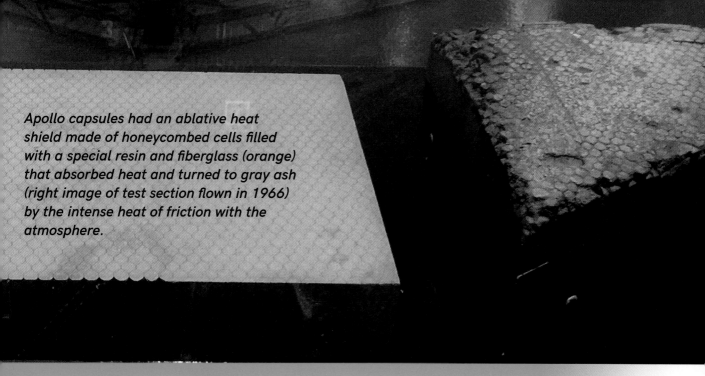

Apollo capsules had an ablative heat shield made of honeycombed cells filled with a special resin and fiberglass (orange) that absorbed heat and turned to gray ash (right image of test section flown in 1966) by the intense heat of friction with the atmosphere.

FACING THE FIRE

Speeding fast enough to reach the Moon is hard and slowing down again is equally difficult. Crew capsules such as the Soyuz TMA, the Starliner, and Dragon 1 are designed to slow down from an orbital speed of 17,000 mph to near zero on the ground. These capsules cannot be used for trips to and from the Moon unless they are fortified with new heat shields. Spacecraft falling from the Moon are going much faster, 24,000 mph, and must slow down that much more.

Re-entry friction produces enough heat to strip electrons from atoms in the atmosphere: producing ions and free electrons—called PLASMA. The Space Shuttle used tiles to protect the crew from this 3000-degree plasma. But those tiles were fragile and expensive to repair after each flight. So the American Orion and the Russian Federation are using heat shields similar to those used on Apollo and Zond capsules.

The Apollo heat shield absorbed the heat and melted, or ablated, taking the heat with it. The ablative material was pushed into a honeycomb structure using a sort of caulk gun. The Orion capsule uses a similar ablative material.

To reduce the load on the heat

FIRST WOMAN ON THE MOON

All the Apollo astronauts were male. No one yet knows when the first woman will walk on the Moon.

The United States is poised to launch Orion atop its SLS booster for a test flight in 2020 (EM-1) and follow with crew on EM-2 about two years later. The crew may include one or more women and will get within 5500 miles of the Moon. NASA then plans to construct its Gateway station starting in 2024. This station would have up to seven modules and serve as a rest-stop and fuel station for lunar missions. Gateway crews may include female NASA and ESA astronauts. Current estimates put an American Moon landing between 2024 and 2028.

The planned SpaceX flight of Maezawa in the mid-2020s will include another un-named person that may be the first woman to see the lunar far side with her own eyes. SpaceX has not announced plans for any landings.

Russia currently has no female cosmonauts, though some may be added by the time Federation flies to the Moon in the mid to late 2020s. Russian human surface landings are not expected until the 2030s.

China is looking to land humans at their south polar research station between 2031 and 2036. They currently have two female TAIKONAUTS.

Female Space Pioneers (counterclockwise): Russian Valentina Tereshkova in 1963; American Sally Ride in 1983; Private spaceflight participant Iranian-American Anousheh Ansari in 2006; and Chinese Liu Yang in 2012. Who will be the first woman on the Moon?

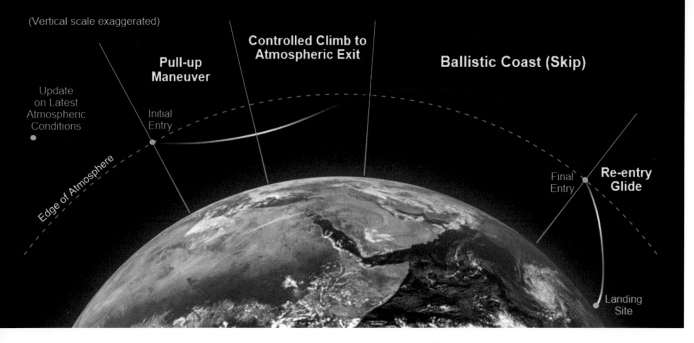

Vehicles returning at high speed from the Moon may use an aerobraking "skip" maneuver into and out of the Earth's atmosphere to slow down before final entry.

shield, the atmosphere of Earth can be used to slow down via a process called AEROBRAKING. Like skipping a stone on water, the vehicle "skips" into the atmosphere and is slowed down by friction with the air. After this "SKIP MANEUVER," the vehicle pops out to cool off and enters again at a lower speed. The number and duration of the "skips" can be adjusted to slow the vehicle down the desired amount without using fuel for braking. The Russians used aerobraking to slow their Zond craft so it could land on land instead of water. The Federation capsules as well as the American Orion plan to take advantage of the fuel savings that aerobraking offers even though these maneuvers add some time and complexity to the entry phase.

STAYING ON THE MOON

New boosters, crew capsules, and landers will soon be ready to launch a new generation of lunar explorers. These early pioneers will scout out the best places to build the first outposts and research stations. Keeping them supplied from Earth will not be practical. How can lunar resources provide habitats, power, water, fuel, and entertainment so people can safely live and work on the Moon?

CHAPTER 5
LONG TERM STAY

Getting to and from the Moon requires energy, mostly in the form of fuel. Staying on the Moon also requires energy, but mostly in the form of electricity. Electricity is needed to power fans, pumps, air conditioning, lights, heaters, robots, radios, computers, motors, and even the toilet. Even more importantly, with enough electric power, lunar pioneers can extract water and minerals to build the first permanent lunar stations. Where are the best places to put these stations?

Astronauts enter a lunar outpost. Orange balls are oxygen tanks wrapped in reflective foil to stay cold.

SAFETY FIRST

The first criteria for choice of Apollo landing sites was crew safety. Trying to land on a steep slope or among rocks would risk disaster. The first landings were in the maria because robotic spacecraft had shown that they were relatively flat and rock-free. New maps and robotic landers will help planners find the safest places.

Another big factor for human landings are the abort options—what to do if there's an Apollo 13 type of explosion or if a lunar module engine fails. Communications and the time available to wait for repair or rescue can mean the difference between life and death. Therefore, spacecraft are sure to be equipped with redundant systems, extra supplies, and well-trained crews backed by experts in Mission Control. But some landing sites have more or different risk than others. For example, far side landings will require more complicated communications than direct signals to Earth used on Apollo. And launch from polar sites may require longer duration transfer orbits than Apollo returns to low lunar orbit from sites near the equator.

One way to increase the safety of early landings is to prepare a "SAFE HAVEN" ahead of time. Like storm cellars, safe havens provide temporary shelter and supplies such as power, food, water, and first aid. A lunar safe haven might be an extra lunar ascent module or a pressurized rover that can be driven robotically to aid a stranded crew.

Whether there is a safe haven or not, all human activities on the surface will require power. The need for power may drive the location of the first human outposts.

A rover with a pressurized cabin such as the one shown at Johnson Space Center in Houston could serve as a mobile home for extended explorations or as a taxi to and from landing sites and early outposts.

Each of the eight International Space Station solar arrays are 112 feet long and 39 feet wide. They use silicon solar cells on golden blankets to capture Sunlight that provides electrical power to the station.

POWER FROM THE SUN

Apollo astronauts used disposable batteries while on the Moon. They only had enough to last for up to three days. Staying for longer periods would require a lot of batteries hauled up on rockets from Earth.

Fortunately, the Moon has lots of sunshine and no clouds to get in the way. Arrays of solar cells can collect this sunshine and convert it into electricity.

The International Space Station has been operating on solar power for more than 15 years. It has eight solar panels that rotate to face the Sun. They produce about 100 kilowatts (kW) of power. That's enough to run 100 thousand-watt microwave ovens for an hour, or 200 of them for a half hour. Waste heat is removed by 14 radiator panels that are 6 by 10 feet long. A similar solar power system could support a base on the Moon—at least during dayspan.

About 60 percent of the station's electricity goes to recharging batteries for nighttime use. But orbital night is only 35 minutes long. Lunar nightspan last two weeks! The ISS needs 24 batteries, each weighing 430 pounds to store the power needed during the night. To provide power for one nightspan on the Moon would require about 8000 batteries weighing more than three MILLION pounds! To get them to the Moon would take an impossible 38 Saturn V launches!

Power for nightspan is one of the main drivers for building the first human outpost at a lunar pole. Both poles have mountains that are in sunlight about 80 percent of the time. Less darkness means a lot fewer batteries and thus fewer launches to build a base.

The poles also are where the most water is located. Electricity can split this water into hydrogen and oxygen during dayspan. These two elements can be recombined inside a fuel cell to generate electricity (and water as a byproduct) during nightspan. Fuel cells are what powered the Apollo CSMs and space shuttles.

A space shuttle fuel cell was about the size of a two-drawer file cabinet and weighed 255 pounds. Three of these supplied an average of 21 kW. About four times that many would provide space station power levels. Tanks to hold the water, hydrogen, and oxygen would also be needed. (These could serve a dual purpose as storage for spacecraft fuel.)

So the Sun can supply the power needs of an early lunar base if that base is located near one of the poles or is only occupied during dayspan. Once a power station is established, it can be scaled up to provide power to other locations via tanker truckloads of fuel, power lines, or wireless transmission via lasers.

Batteries on the space station (white boxes on truss where white radiator panels descend from below the golden arrays) are designed to last about ten years. The crew used the Canadian robotic arm (shown) to swap out some old batteries for new ones in 2017.

Artist's concept of a 10-kilowatt "kilopower" reactor on the Moon. Umbrella structure is a radiator to cool the reactor.

GOING NUCLEAR

Another way to solve the lunar power problem is to go nuclear. A nuclear reactor provides the same amount of power day and night. No batteries needed. No limits on location!

NASA and the U.S. Department of Energy are testing small portable reactors now. One 10-kW "kilopower" unit would weigh about 3300 pounds. Ten of these could replace the power provided by the space station's solar arrays and batteries. The weight is low enough for one Saturn V to put them on the Moon.

Each unit contains a coffee can-sized core that holds the fuel. The fuel is uranium 235 (U-235) that splits or fissions when hit by atomic particles called neutrons. This fission releases heat and more neutrons. These neutrons strike other U-235 atoms and keep the process going. The heat is used to drive a piston that produces electricity, much like burning gas does in a car engine. An umbrella-shaped radiator cools the reactor.

Neutron radiation can cause cancer. So during transport or refueling, a rod (of boron carbide) is inserted in the core to stop the chain reaction of neutrons. Placing a reactor in a crater about two thirds of a mile away or shielding it with water or concrete will keep people safe.

After about ten years, the amount of U-235 remaining is not enough to keep the chain reaction going. The reactor shuts down. The used core will still be radioactive. Robots will likely take it to a disposal crater away from people. A new core is needed to restart the reactor.

The Moon has uranium but turning it into fuel is very complex and regulated. So U-235 would likely be imported from Earth, or reactors might be reconfigured to use THORIUM. Thorium fuel can be recycled and is not suitable for weapons. The western near side has the most thorium.

A different kind of nuclear fuel is found on the Moon that rarely occurs on Earth: HELIUM 3. When struck by a neutron, helium 3 becomes helium 4 and releases energy. No neutrons are released, so there is no radioactive waste. So far, scientists have not found a way to keep the reaction going. But if/when they do, helium reactors may become a source of clean energy on Earth. A slow-rolling tractor could remove helium 3 from regolith by simply heating it.

Nuclear power is an option for any location on the Moon. Only one launch may be required to set it up. Refueling from Earth or lunar resources will be required after 10 years.

WATER: FUEL FOR LIFE

Water is essential to life. It's also useful for washing and cooking and as a radiation shield. But in space, the most important thing about water is that it can be split into hydrogen and oxygen. Of these, oxygen is by far the most important.

About 85 percent of a spacecraft's mass at launch is oxygen used for fuel. To make electricity, fuel cells use about eight pounds of oxygen for every pound of hydrogen. Plus, an average person consumes 1.8 pounds of oxygen per day.

About 65 percent of the water on the ISS is recycled from air and from crew waste. The rest needs to be resupplied. Because of the Moon's lower gravity, water from the Moon to LEO takes much less energy to deliver than it does from Earth.

The polar areas of the Moon have the most water. Some of this water is deep in

WELCOME TO THE MOON

craters cold enough to turn oxygen to ice. Designing equipment to work in this extreme cold poses a challenge. But these deep dark "refrigerators" may also offer natural storage places for fuel.

Oxygen can also be extracted from Moon rocks using a process called HYDROGEN REDUCTION. ILMENITE—a mineral made of iron, titanium, and oxygen—is ground into sand and heated. Then hydrogen gas is bubbled through the sand. Oxygen grabs onto the hydrogen and forms water. Electricity removes the oxygen from the water. The hydrogen is recycled. Iron and titanium are also produced in the process. These metals are useful for making magnets, paint, and spacecraft parts.

Lunar water and oxygen offer enormous savings in transportation costs and security against accidents. Selling excess supplies to space stations may also jumpstart the first lunar businesses.

THE RECYCLING TOILET
The Russian waste collection system (toilet is white lid with hose extended) splits waste water into hydrogen and oxygen. The oxygen goes back into the cabin. The hydrogen is mixed with carbon dioxide from the air and turned into methane and water. The methane is vented overboard, and the water is reused. The American toilet (not shown) recycles urine into clean water. A separate system collects water from the air for reuse.

This image of the Sea of Cleverness is famous for its light-colored swirls ("S") associated with magnetic "hot spots." The arrow shows the location of a lunar sky light that is 427' wide and 1805' deep. Sky lights are assumed to form when lava tubes collapse. Recent data suggest lava tubes up to a half mile wide lurk under the lunar surface.

THE LUNAR UNDERGROUND

At sunset on the Moon, the temperature can drop from 260°F to -280°F. Temperature shifts like this are hard on equipment. Metals expand and contract, liquids freeze and thaw, and plastics become brittle. Lunar pioneers will need to keep repair kits (including 3-D printers) and spare parts handy.

The thermal "climate" at the poles is much more mild. With the Sun always on the horizon, the surface temperature is a constant -58°F. Elsewhere, pioneers can escape underground. Data from Apollo shows that just six feet down, the temperature holds steady around zero degrees. The use of natural underground tunnels, called LAVA TUBES, some up to a half-mile wide, may save on construction costs.

Going underground is also a good way to avoid radiation. Radiation comes from the Sun in the form of x-rays and protons. Other stars add fast-moving particles

WELCOME TO THE MOON

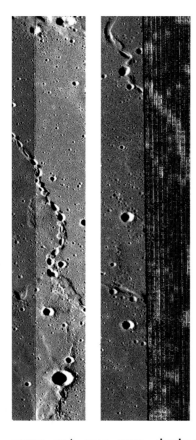

Lava tubes may provide ready-made habitats for humans in the future. But except for skylights, they aren't visible on the surface. Because lava tubes are warmer during nightspan and cooler during dayspan than other areas, thermal data can reveal their locations.

called galactic cosmic rays.

Radiation is measured in rems. One REM is equal to about 670 dental x-rays or one full-body CT scan. The Apollo 14 crew got one rem during their six-day mission. Most of that dose came from passing through the VAN ALLEN RADIATION BELTS. These belts are electrically charged particles trapped by Earth's magnetic field. Only five percent of the Apollo 14 dose came from time on the Moon.

Apollo astronauts "saw" particles of radiation as "jet trails" in the fluid of their eyes. They also got small burns on their skin from particle impacts. These effects caused no permanent damage.

However, radiation can have hidden effects. Damage may show up years later as cataracts or cancer. An accumulated dose of 400 rem is predicted to increase a person's risk of cancer by 3 percent. American astronauts are grounded when they accumulate this dose.

The lunar surface gets 70 to 130 rem per year. But under 20" of regolith, the dose drops to 20 rem per year. People could work on the surface for four years or underground for 20 years. These times may be extended as scientists learn how to prevent radiation damage with special clothing, diets and drugs.

People also need protection from solar flares. Big flares come in cycles about every 11 years. One flare can produce a year's worth of radiation in a few days. A sudden dose of 100 rem causes vomiting and hair loss. Only half of the people exposed to 300 rem survive. Flares produce x-rays and particles. X-rays travel at the speed of light. Particles are much slower. So x-ray detectors can provide several hours warning, giving people time to reach safe havens.

Mizuna (mustard greens) grew in nutrient-rich water under artificial light on the International Space Station in 2009. Analysis of the plants showed some DNA damage from being in space.

GREENING THE MOON

Apollo crews brought all their food from Earth. But like using batteries for power, that's not a good long-term solution. So as soon as power and water are available, a lunar base is likely to add a greenhouse.

Plants need NUTRIENTS (elements) to grow. On Earth, all but three (carbon, oxygen, and hydrogen) come from soil. But lunar regolith lacks organic waste, microbes, and nitrogen found in Earth soil. It is also low in potassium and phosphorus and may have toxic levels of nickel and chromium.

It should be possible to manufacture synthetic soil out of lunar regolith. But early lunar greenhouses are likely to use HYDROPONICS—growing plants in lunar water with nutrient packages from Earth. Similar systems have produced lettuce eaten by astronauts on the International Space Station. Also 10 to 20 times more plants can be grown in a container of water than in soil because plants don't have to compete with each other for water and minerals.

Yams and wheat may be among the first crops grown on the Moon. Both grow well hydroponically and are loaded with vitamins and minerals people need. Eighty percent of the yam plant is edible.

Whereas yams need some darkness, wheat can take constant light. Wheat plants developed for space are also short and fast growing. They take up very little space and can be ready to eat in less than a month.

Plants provide psychological benefits and freshen the air. They soak up carbon dioxide and release oxygen. In a NASA test, a man was locked in a chamber for 30 days with 30,000 wheat plants. They soaked up all of his carbon dioxide and provided all the oxygen he needed.

Lunar farmers can grow food using lunar water with nutrient packages and seeds imported from Earth. But many experiments will be needed to find the best ways to grow crops in the low gravity, high radiation, and lighting conditions of the Moon.

FLYING TO STAY HEALTHY

So far only twelve men have walked on the Moon. Those trips didn't appear to cause any permanent damage to their health. But no one yet knows if the low gravity of the Moon is enough to keep humans healthy for extended stays.

Studies show that exercise is key to good health. On Earth, walking requires effort from muscles and bones, including the heart. Floating in space is akin to lying in bed all day. To prevent muscles and bones becoming weak from lack of use, ISS astronauts spend hours every day running on treadmills with straps to hold them down. Lunar crews can do that, too. But, they can also fly to stay healthy! A 100-pound person only weighs 17 pounds on the Moon.

So, once a large area in enclosed with air, strapping on wings and flapping should lift them right off the ground.

Falling won't be as much of an issue, either. On Earth, a fall from a two-story (18-foot) building could be deadly. The fall takes about a second, and the impact speed is about 22 mph. On the Moon, it would feel like a six-foot drop. The fall lasts three seconds and the impact speed is only 9 mph. Imagine what a high diver could do with all that extra time!

The ability to jump higher and the slow pace of falling will require traditional sports to be adjusted for the Moon. Basketball hoops may be raised. Races may take longer.

During Apollo 16, John Young took advantage of the Moon's low gravity to jump several feet off the surface while saluting the flag.

WELCOME TO THE MOON

Lunar pioneers may hold monthly "Terminator" races. The day/night line moves across the surface at about 10 mph. With the low gravity, an athlete could easily run that fast. Marathons might chase the Sun all the way around the equator during a lunar rotation.

Unusual sports might form the basis for an early lunar business. People on Earth may pay to watch lunar dune buggy races or golf tournaments. Some adventure tourists may pay to climb the slippery slopes of Malapert Mountain that rises nearly three miles above the lunar south pole. Others might prefer spelunking lunar lava tubes in the Marius Hills on the western limb. Maybe there's a chamber of secrets waiting to be discovered?

WORKING ON THE MOON

Working on the Moon requires constant awareness of the environment, especially when moving around. Apollo spacecraft and rovers sprayed dust everywhere. Astronauts and tools were covered with it after their spacewalks.

On Earth dust grains become rounded by being rolled around by wind and water. Moon dust made from impacts are like tiny shards of glass. These bits of grit lodge in cracks and act like sandpaper to wear down moving parts in space suits, rovers, and drills.

Apollo capsules did not have airlocks. Astronauts tracked lots of dust into their modules. They reported it smelled like damp ashes. The crews did not get sick from the dust. But breathing dust long term, like air pollution, is not good for the lungs.

To keep Moon dust outside, crews might use spacesuits with a "turtle-back" design. The wearer backs up to an airlock. The "shell" opens like a small hatch, and the astronaut climbs out. But supplies and equipment will still require full-sized airlocks. Because lunar dust is high in iron, airlocks might be outfitted with magnets to remove dust. Filters will then remove any that gets into the air.

Modules will also be equipped with vacuum cleaners. Like a hose on a home vacuum, dust and other waste will be sucked out through a tube. But some air is lost with each use. So preventing dust from getting onto equipment in the first place is the best solution.

Work areas may be tramped down or cleared with shovels or robotic rovers. Trails to power stations or landing sites

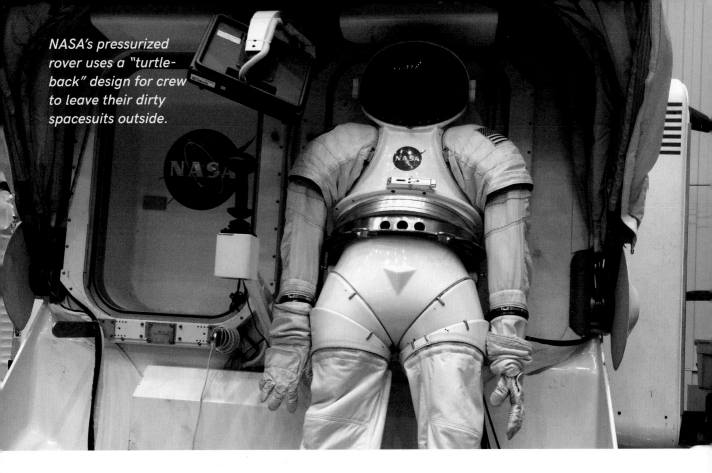

NASA's pressurized rover uses a "turtle-back" design for crew to leave their dirty spacesuits outside.

may be overlaid with raised metal "boardwalks" and eventually paved. Regolith has been shown to melt into glass in a kitchen microwave. So sidewalks and roads may be "paved" to glass by rolling a hot drum over them.

For longer distances, or to cross craters or between mountains, pioneers might put in zip lines. These cables may be made of lunar metals such as iron and titanium that are byproducts of extracting oxygen from lunar rocks.

As lunar mining operations increase, and fuel is available, lunar workers may commute via rocket-powered chairs strapped to their spacesuits or in some sort of flying rover. Landing pads inside of craters will keep the dust from escaping onto nearby solar arrays.

Staying on the Moon will not be easy. Early stations will need power, water, air, fuel, radiation protection, and food. Lunar resources can help with all of these things, most easily at the lunar poles. The Moon also offers rocket fuel, helium 3, water, and metals that may be exported to Earth, space stations, and eventually Mars.

CHAPTER 6
HEADING OUT

The benefit of learning how to live away from Earth will allow humans to spread life to more worlds—with Mars the next obvious destination. Though Mars is different from the Moon, staying on either world requires power and life support systems that will work for years despite extreme temperatures, vacuum, radiation, and dust. Testing these systems in Earth's "back yard" will build confidence that humans can safely venture further into the solar system.

The surface of the Moon is equal to the land area of Africa and Australia put together. Its water and minerals are a treasure of raw materials to spur new business and opportunities. Understanding its evolution will help scientists locate similar resources on other planets and moons. So why hasn't anyone been back to the Moon since the days of Apollo?

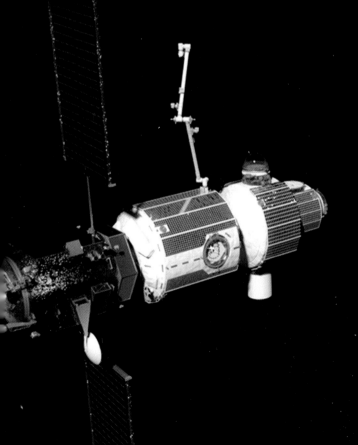

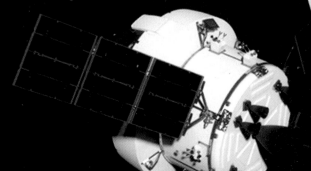

NASA plans to place a Gateway in lunar orbit to serve as a fuel depot, communications relay, and staging area for missions to the Moon and Mars. A power and propulsion element (left with solar arrays) is targeted for launch in 2022. A robotic arm, habitation module, and airlock (left white modules) may be added starting in 2024. Spacecraft (right) will deliver crew and cargo.

THE VALUE OF THE MOON

The reason most often quoted for why no one has returned is that it is too expensive. Why should people risk their lives and spend billions of dollars to go to the Moon?

The Apollo Program took about eight years and cost about $144 billion (in 2017 dollars). The International Space Station assembly took about eleven years and cost about $150 billion.

What value was Apollo? Some analysts say that every dollar spent returned eight to ten dollars in economic benefit. For example, computers in the 1960s weighed tons, filled rooms, and needed their own air-conditioning systems. To reduce launch mass, NASA shrunk the Apollo guidance computer to a 70-pound rectangle that would fit in carry-on luggage. Today's lightweight computers were driven by the needs of Apollo. Other spinoffs include remote health monitoring, fire resistant materials, and even improved baby food.

Yet the scientific and technical spinoffs of Apollo pale in comparison to its less-visible benefits. The space race avoided nuclear war, and the Moon landing inspired a generation with a can-do spirit. What are these benefits worth?

The ISS also has its share of spinoffs. But its most long-lasting benefit may be the proof that people from different cultures, even former enemies, can successfully live and work in space together. Men and women have performed more than 200 spacewalks to assemble, replace, and repair station equipment. Cargo vessels from the United States, Russia, Japan, and European nations have successfully delivered supplies year after year. Scientists and students around the world share in the excitement of new discoveries.

People are sure to say building an outpost on the Moon is too expensive. Many of them don't realize that NASA's budget is only a half cent of every dollar the government spends. In partnerships with other nations, this level of funding was enough to fly the space shuttle, build the ISS, and support robotic missions to other planets. And it is enough to continue ongoing programs while developing new spacecraft and systems needed to reach the Moon—this time to stay!

MOONBASE ONE

Imagine the awe of viewing a mountain taller than any in the United States with your home planet glowing on the horizon. Under the mountain's dusty gray blanket, lies a storehouse of elements and minerals, building blocks for a new frontier. The Moon's low gravity and natural lack of air are ideal for creating beautiful glass, strong alloys, and critical fuels. In deep and mysterious craters, ice and exotic elements like helium 3 also offer opportunities for new industries to sustain human settlements.

The stark and alien landscape also attracts tourists and artists intent on experiencing and capturing its magnificent beauty. Robotic scouts are busy updating maps to pinpoint exciting new places to visit such as the Mountain of Eternal Light—a place near the South Lunar Pole that is nearly always in sunlight.

Already, pioneers in science and business are making plans to send spacecraft and build settlements on the Moon. They will produce fuel from abundant oxygen in the soil and build spacecraft parts from rocks rich in titanium. They will show tourists how to ski on slippery lunar dust and prove to them that the Apollo 11 flag really did fall over!

Life on the lunar frontier will not be easy. Solar flares and meteor hits are a constant threat. Two-week nightspans take some getting used to. But humans are finding ways to grow food, pull fuel from rocks, and harness the power of the Sun. They are inventing low-gravity sports to play in the lunar underground. Perhaps you will be one of these people, a pioneer who will reap the bountiful harvest of the Moon and call it home.

Moonbase by Alex Aurichio, winner of a National Space Society student art contest.

UNITS CONVERSION TABLE

One English Unit	Multiplied Times	Equals International Units
Foot (ft x 12 = inches)	0.3	Meters (m)
Miles (mi)	1.6	Kilometers (km)
Square Mile (sq. mi. or mi^2)	2.59	Square Kilometers (sq. km or km^2)
Pound (lb)	0.45	Kilograms (kg)
Ounce (oz x 16 = pounds)	28	Grams (g x 1000 = kg)
Ton (t x 2000 = pounds)	0.9	Metric Tons (mt)
Gallon	3.8	Liters (l)
*Fahrenheit	5/9 x (F-32) + 273	°Celsius (C) (°C-273=°Kelvin [K])

MOON FACTS

REGOLITH COMPOSITION
42% oxygen, 21% silicon, 13% iron, 7% aluminum, 6% magnesium, 3% other

DIAMETER
27% of Earth's, 2,160 miles (3,474 km)

MASS
1% of Earth, 3.3 x 10^{22} pounds (7.3 x 10^{22} kg)

SURFACE AREA
7% area of Earth, 14,600,000 square miles (37,800,000 sq. km)
land area of Africa plus Australia

GRAVITY
1/6th of Earth (0.17g), 5.3 ft/s^2 (1.6 m/s^2)

DISTANCE
226,000 mi (closest), 239,000 mi (average, 385,000 km), 252,000 mi (farthest)

TEMPERATURE EXTREMES
south pole -387°F (coldest), equator at noon, 243°F (warmest)

AVERAGE TEMPERATURES
225°F (dayspan), -243°F (nightspan)

ROTATION (time to spin on its axis)/REVOLUTION (time to orbit Earth)
27 days 8 hours (656 hours)

SYONDIC PERIOD (time between full moons)
29 days 13 hours (709 hours)

ROCKS BROUGHT BACK BY APOLLO
842 pounds (381 kg)

FOR MORE INFORMATION

To order a 25-foot Giant Moon Map (back cover) for a school or event, visit
ShareSpace.org
To schedule a school visit, contact the author at MarianneDyson.com

GLOSSARY

ABLATED – burned off. Ablative materials on spacecraft heat shields absorb heat during entry

ABORT – end/turnaround/discontinue a flight or mission because of a spacecraft problem or failure

AEROBRAKING – using the atmosphere of a planet as a "brake" to slow down without using fuel

ALBEDO – brightness or amount of light reflected off a body such as the Moon. White objects have a high albedo and black objects a low albedo

ANORTHOSITE – rock made mostly of calcium, aluminum, silicon, and oxygen that is white to light gray. It formed the original crust of the Moon

APOLLO – a project of the United States of America that took men to the Moon between 1969 and 1972

BASALT – dark-colored rock rich in iron and magnesium, created by the solidification of lava. Lunar basalts are found in the maria

BRECCIA – rock made of fragments of other rocks as the result of impacts

CIRCUMLUNAR – around the Moon

COMMAND AND SERVICE MODULE (CSM) – Apollo spacecraft that took men to lunar orbit and back to Earth.

COSMIC RAYS – fast-moving charged particles of radiation from outside the solar system.

CRYOGENIC – extremely low temperature substances. Liquid oxygen and hydrogen are cryogenic fuels.

DAYSPAN – daytime on the Moon. A dayspan lasts two Earth weeks

DELTA V – a measure of acceleration or change in velocity that indicates how much fuel is required to change direction/orbits in space

EJECTA – material such as rocks and lava thrown out as a result of a volcanic eruption, meteorite impact, or explosion

ESCAPE VELOCITY – the lowest velocity required to overcome the gravitational attraction of a planet or Moon

EXOSPHERE – the Moon's atmosphere, a thin layer of gas and dust between the surface and deep space

FAR SIDE – the side of the Moon that faces away from Earth at all times. It is not dark. It receives as much sunlight as the near side.

FREEFALL – downward movement caused by the force of gravity. All masses fall at the same rate and thus appear to be floating inside spacecraft

FUEL CELLS – rechargeable batteries that combine hydrogen and oxygen to make electricity and water. Fuel cells may provide power for rovers and during the lunar nightspan.

GIANT IMPACT THEORY – says the Moon formed as the result of an impact

GRAVITY – the force that attracts bodies toward each other. Earth's gravity (g) is $g=GM/R^2$ where G is the gravitational constant, M is mass of Earth, and R is radius of Earth.

HEAT SHIELD – a coating or device to protect a spacecraft from high temperatures during atmospheric entry

HELIUM 3 – a form of helium that has one instead of two neutrons in the nucleus. It may be used for nuclear fusion

HYDROGEN REDUCTION – a method of removing oxygen from lunar rocks using hydrogen gas

ILMENITE – an iron-titanium mineral found in mare basalts that can be used as a source of oxygen

ISOTOPE – a form of an element such as helium or uranium that has a different number of neutrons

LEO – (pronounced "lee-oh"), an acronym for low Earth orbit which is up to 1200 miles (2000 km) above the surface of the Earth

LIQUID OXYGEN (LOX) – (pronounced "locks") an acronym for liquid oxygen that is made by chilling oxygen gas to cryogenic temperatures. LOX is used in rocket engines and fuel cells

GLOSSARY

LIQUID HYDROGEN (LH2) – (pronounced "el H two") Liquid hydrogen is made by chilling hydrogen gas to cryogenic temperatures. LH2 is used in rocket engines and fuel cells.

LUNAR MODULE – (LM, pronounced "lem") an Apollo spacecraft that landed on the Moon. After each mission, these modules crashed on the Moon

LUNAR ORBIT INSERTION – a braking maneuver that allows a spacecraft to enter orbit around the Moon

MAGMA – hot fluid material under the surface of a planet or moon which forms lava during eruptions and basaltic rock when cooled

MANTLE – an interior region of a planet or moon that is between the crust and the core

MARIA/MARE – the Latin word for sea. Maria formed when dark lava flowed into low places

MASCON – a concentration of dense material below the surface of a moon or other body causing a local increase in gravitational pull and, depending on composition and formation, an increase in magnetism

NEAR SIDE – the side of the Moon that always faces Earth

NIGHTSPAN – night time on the Moon. A nightspan lasts two Earth weeks.

PAYLOAD – the amount of goods or materials that a spacecraft can lift to orbit. Payload is generally less than 15 percent of the total mass of a rocket at liftoff.

PLASMA – a gas consisting of atoms (ions) missing one or more electrons and electrons not bound to atoms. Earth's outer atmosphere contains plasma. The Sun and solar wind are made of plasma.

RADAR – a system that bounces high-frequency waves off objects to measure distance, direction, and speed

REGOLITH – loose rock and mineral fragments caused by impacts. The surface of the Moon is covered by it.

REM – a unit for measuring radiation. One rem is equal to the radiation in about 670 dental x-rays

RENDEZVOUS – a meeting of two objects or spacecraft in space that requires them to match velocities

RIFT VALLEY – a long channel with steep walls formed when the crust of the Moon spread apart

ROCKET PROPELLAND-1 (RP-1) – is refined kerosene, derived from petroleum and stable at room temperature. RP-1 was used for the first stage of the Saturn V and also modern rockets.

SATURN V – a powerful booster rocket used by the Apollo program to send men to the Moon.

SEISMOMETER – a device that measures vibrations or movements of the ground caused by impacts or quakes.

SOLAR WIND – charged particles from the Sun that move at about a million miles per hour.

SOLDER – a metal alloy that is melted to join two pieces of metal together.

SUBLIMATE – when a solid changes phase directly into gas, instead of into a liquid and then a gas

SYNCHRONOUS ROTATION – a rotation period that equals the period of revolution so that one side of a body faces the star, planet, or moon that it is orbiting. The Moon is in synchronous rotation around Earth because of tidal forces.

TAIKONAUTS – Chinese astronauts

TERMINATOR – the dividing line between sunlit and nightside of a moon or planet

THORIUM – a radioactive metal element that can be used as fuel for nuclear fission. Lunar thorium is found around the Sea of Rains

TRANSLUNAR INJECTION – a maneuver used to propel a spacecraft fast enough (7 miles/sec) to reach the Moon

TITANIUM – a gray metal element used in spacecraft that stays strong at high temperatures

VAN ALLEN RADIATION BELTS – two zones around Earth where electrically charged particles are trapped in Earth's magnetic field.

ZIRCON – a mineral used to date rocks. Fresh zircon contains uranium but no lead. Uranium decays into lead at a known rate. Lead in a zircon crystal therefore reveals the age of a rock.

INDEX

A
Aldrin, Andrew, 5, Buzz, *see also* Apollo 11, 5, 21, **29**, **32**, 33
Alphonsus Crater, 24, **25** (Ranger in)
ALSEP, 34, 36, **41**,
Anders, Bill, *see also* Apollo 8, **28**
Anousheh Ansari, **72**
Antares (spacecraft), 36 (Apollo), 66
Apollo 1, **28**, 31 (timeline)
Apollo 7, 8, 9, 10 missions, 30, 31
Apollo 11, 5, **20**, **25**, 31 (timeline), **32**, 47, **63**, **64**, 90
Apollo 12, **12**, 27, 31 (timeline), 34, 36, 46, 47
Apollo 13, **35**, 37 (timeline), 40, 75
Apollo 14, 11, **25**, **36**, **37** (timeline), 46, 47, 82
Apollo 15, **9**, **14**, **16**, 37 (timeline), 38, **39**, 46, 47, 48
Apollo 16, 37 (timeline), **40**, **41**, 44, **46**, 47, 55, **85**
Apollo 17, 37 (timeline), **42-43**, 44, 46, 47
Apollo command module (CM) *see* command and service module
Apollo lunar module (LM), *see* lunar module
Aquarius (spacecraft), 35
Aristarchus Crater, 11, **82**
Armstrong, Neil, *see also* Apollo 11, 21, **32**, 33

B
Bay of Rainbows, 58
Bean, Alan, *see also* Apollo 12, **12**
Beresheet(spacecraft), 59
Bezos, Jeff, 65
Boeing (company) *see* CST-100
Borman, Frank, *see also* Apollo 8, 30

C
Cabeus Crater, **53**, 54
capsule, *see also* command module, 65, 66, **68** (Orion), 70, 71, 73
carbon dioxide, **35** (filter), 68, 80 (toilet recycling), 84 (plants)
Casper (spacecraft), 40
Central Bay, 12 (location of), **19** (red cross)
Cernan, Gene, *see also* Apollo 10 and 17, 31, **42**
Chaffee, Roger, *see also* Apollo 1, **28**
Chandrayaan-1 (spacecraft), 52, 54
Chang'E (spacecraft) *see* China/Chinese
China/Chinese, 5, 52, **58-59** (Chang'E), **60** (Long March), 65-67, 70, **72** (Liu Yang)
Clementine (spacecraft), **49**, 50, **51**, 52
CM, *see* command and service module
Collins, Michael, *see also* Apollo 11, 21, 32, 33
Columbia (spacecraft), 21, 32
command and service module (spacecraft), *see also* capsule, **16**, 19, 21 (Columbia), 22 (Gumdrop), 28, 30, **31**, 35 (Odyssey), 36 (Kitty Hawk), 40 (Casper), 42 (Falcon), 63, 64 (Columbia), 66, 68, 77, 93
Cone Crater, 36, **37**
Conrad, Pete, *see also* Apollo 12, **34**
Copernicus Crater, **18**, 27, **47**
core, 8-9, 34, 40 (drill core), 43 (core sample), 46, 48, 94
crust, 8-10 (formation), **11** (thin section), 38, 46, 48, **55** (lunar thickness map), 56, 93-94
Cosmic rays, *see also* Radiation, 48, 82, 93
crater, **10** (Crises), 11, **18** (Copernicus, Grimaldi, Tycho), **19** (Tsiolkovsky), 23, 24, **25** (Alphonsus), **26** (Tycho), 27, 36, **37** (Cone), 38, 40, 42, **43** (Shorty), 44-45, **47** (Copernicus), **49** (Plaskett), 50 (hydrogen in), 51-52, **53-54** (Cabeus), 55-56, **57-59** (Tycho glow, Von Kármán), 78, 80, **82** (Aristarchus) 87 (landing pads inside), 90
CSM, *see* command and service module
CST-100 (Starliner, spacecraft), 65

D
Descartes Crater, 40
Delta IV (rocket), 65, 66, 68
Duke, Charlie, *see also* Apollo 16, 40

E
Eagle (spacecraft), 21, 32, 33
Eastern Sea, **18**
Europe and European Space Agency, 5, 52, 65, 66, 70, 89
experiment, 28-29 (Luna), 33-34, 36, **41** (ALSEP), 52 (radar), 62, 63, 84
Exploration Mission, 68
ESA *see* European Space Agency
Eyles, Don, 36

F
Falcon (spacecraft), 42 (Apollo 17), **65** (SpaceX), 66, **67** (reusable), 70
far side, 12, **13** (day/night on), **18-19** (features labeled), 23-24, 30-31, 33, 36, 38, 48, **51** (Moscow Sea), **55** (crustal thickness map), 56, 59, 72, 75, 93
Federation (spacecraft), 68, 70-73
food, *see also* plants, 62, 75, 83, 84, 87, 89, 90
freefall, 17, 93
fuel, 19, 21 (Apollo 11), 32, 33, 40, 61, 62, 66, 67 (types), 69, 72 (station), 73 (for braking), 74, 77 (fuel cells), 78-79 (nuclear), 79, 80, 87 (mining), 88 (fuel depot), 90, 93-94

G
Gagarin, Yuri, 24
Gateway (space station), 67, 69, 72, **88**
Giant Impact Theory, 8, 93
Gordon, Dick, *see also* Apollo 12, 34
GRAIL (spacecraft), 48, 55 (map), 56, 58
gravity, 15-18, 21, 43, 46, 49, 55-56, 79, 84, **85** (John Young demonstrates), 86, 90, 92 (Earth/Moon values)
Gravity Recovery and Interior Laboratory, *see* GRAIL
Grimaldi Crater, **18**
Grissom, Gus, *see also* Apollo 1, **28**
Gumdrop (spacecraft), **22**

H
H2 (rocket), 66
Haise, Fred, *see also* Apollo 13, 35
Hiten(spacecraft), 52, 58
Hydrogen, 50, 61 (liquid), 67, 77, 79 (from water), 80 (from rocks/recycling), 83, 93-94

I
impacts, 7-8 (formed the Moon), **10**, 11, 19 (spacecraft), 23, 24 (of Rangers), 26 (of Luna),34 (Apollo measured), 36, 40 (history), **43** (created Shorty Crater), 46 (rock timeline), 48, 50, **52** (of Kaguya), **53-54** (into Cabeus Crater), 55, 56 (SPA basin forming), 59, 82 (of radiation), 84, 87 (dust), 93-94
India, *see also* Chandrayaan, 5, 52, 66 (Launch Vehicle)
International Space Station, **76** (arrays), 79 (ISS recycling), 83 (plants grown on), 84 (astronauts), 89 (cost)
Irwin, Jim, *see also* Apollo 15, 38, **39**
Israel, 59
ISS, *see also* space station, 50, 79 (recycling on), 84 (exercise on), 85 (spinoffs),

J
Japan/Japanese, 5, 52 (Kaguya), 65, 66, 70 (billionaire Yusaku Maezawa), 89

K
Kaguya spacecraft), **52**, 55 (map by), 58
Kennedy, John F. (JFK), 23 (timeline), **24**, 63 (space center named after)
Kitty Hawk (spacecraft), 36
Known Sea, 24- **25**
Komarov, Vladimir, *see also* Soyuz 1, 28
Kosmos (spacecraft), 28, 31 (timeline), 44

L
L1 (spacecraft), 28, 30, 21 (timeline), 44
Late Heavy Bombardment (Lunar Cataclysm), 46, 48
LeMonnier Crater, 44
LEO, 16 (definition), 44, 66 (pounds to), 70 (crew to), 79, 93
LH2, liquid hydrogen, *see* hydrogen
Liu Yang, **72**
Long March (rocket), 60 (Long March 5), 65, 66-67, 70
Lovell, Jim, *see also* Apollo 8 and Apollo 13, 30, 35
LRO, *see* Lunar Reconnaissance Orbiter
Luna (spacecraft): Luna 1-3, 23 (1-8 timeline); Luna 4, 24; Luna 5-9, 26, **27**; Luna 13, 27, 29;
Luna 15, 31 (9-15 timeline); Luna 16-24, 37 (timeline); Luna 17-24, **44-45**, 58; Luna 25-28, 70
Lunar Crater Observation and Sensing Satellite (LCROSS spacecraft), 53-54 (crash site), 58
Lunar Atmosphere and Dust Explorer (LADEE spacecraft), 57, 58 (impact)
Lunar module, 19, 21 (Eagle), 30, 31, 32-33 (Eagle), **34**, 35 (Aquarius), 36 (Antares), 40 (Orion), 44 (Russian), 66, **69** (Lockheed concept), 70, 75, 95
Lunar Orbiter 1-5 (spacecraft), 31 (timeline)
Lunar Prospector (spacecraft), 50, 58

WELCOME TO THE MOON

Lunar Reconnaissance Orbiter (spacecraft), 4, 52, 53 (map by)
Lunokhod (rover), 37 (timeline), **44** (with Luna 17)

M

Maezawa, Yusaku, 70, 72
Malapert Mountain, 86
map, 23, **42** (as fender), 47 (link to animated), 49-**50** (of water), **51** (of minerals), 52-**53** (of temperatures), 55 (of crust), 56, 75, 90
mare/maria-*see* sea/Seas
Marius Hills, 86
Mars, 7, 8, 43, 44, 50, 87, 88
Mattingly, Ken, *see also* Apollo 16, 40
McDivitt, James, *see also* Apollo 9, 31
Meteor Crater, 23
Mission Control, 18, 31, 33, 35, 36, **42** (Apollo 13), 75
MIT, 36
Mitchell, Ed, *see also* Apollo 16, 36
Moon: craters on, *see* craters; diameter of, 92; distance, 7, **14** (measuring), 30, 37 (judging), 45 (measuring), 92; equator, 27, 40, 47, 49, 75, 86, 92; far side of, *see* far side; mapping of, *see* map; mare/maria of *see* seas; mountains on, *see* mountains; near side of, *see* near side; resources of, *see* resources; temperature of, *see* temperatures; near side of, *see* near side; surface, *see* surface; temperatures on, *see* temperatures
Mons Rümker (mountain), 59
Moscow Sea, **19**, **51** (minerals of)
mountains, **9** (Hadley), 10-11, 38, 40, 56, 77, 87
Musk, Elon, 65

N

N-1 (rocket), 30-31 (timeline)
NASA, 4, 30, 35, 36, 38, 40, 45, 48, 50, 55, 57, 67, 69 (female astronaut), 70, 72, 78, 84, **87** (rover), 88-89 (budget)
near side, 12 (view from), **13** (phases of), 15 (why faces Earth), **18-19** (features of), 23, **25** (Ranger images of), 48, 52, **55** (crustal map of), 59, 79, 93-94
New Glenn (spacecraft), 65
New Origin (company), 65
Nixon, Richard, 33
nuclear, 41 (ALSEP battery), 78 (power), 79 (fuel), 89, 93

O

Ocean of Storms, **18** (map location), 26 (Luna landings on), 27 (Surveyor and Luna 13 landings on), 48 (rift valleys around), 56 (opposite to SPA)
Odyssey (spacecraft), 35
orbit, *see also* LEO, 12, 13, **15** (diagram), 16-19 (lunar), 30-33 (Apollo in), 36, 38, 40, 42, 43-44-45, 49, 55, 57, 58, 59, 62-65, 68, **69** (lunar module to), 70-71, 75, **88** (Gateway in), 93-94
Orion *see also* capsule, 40 (Apollo LM), **66** (rocket to lift it), **68** (mockup and dimensions), 71-73
oxygen, 30, 35, **74** (at lunar outpost), 77 (at poles), 79-**80** (from water and rocks, recycling of), 84 (from plants), 87, 90, 92 (percent of regolith), 93

P

plants, 28 (sent to Moon), 83 (grown on ISS), 84
Plaskett Crater, **49**
Pioneer 4 (spacecraft), **23**
Power, 24, 35, 45 (radioactive), 62-63 (of fuel cells/LM), 67 (of fuel), 73-**76** (of ISS), 77 (from fuel cells), 78-79 (nuclear), 83, 87-**88** (of Gateway), 90, 93-94

R

radiation, *see also* cosmic rays, 27, 28, 38, 78, 79, 81, 82, 84, 87, 88, 93, 94
Ranger 1- 9 (spacecraft), **23** (timeline), 24, **25** (photos from)
Reiner Crater, 26
rendezvous, 31, 33 (Apollo 11), 63, 65, 70, 94
resources, 8, **49** (of lunar poles), 59-60, 63, 79, 87-88
Ride, Sally, **72**
rocket, *see also* Saturn V, 16, 26, 29 (penetrator), 31, 36 (exhaust of), 46, 53, **60** (Chinese), 61-62 (stages of), **65** (Delta IV, Falcon, Proton M), **66** (solid, currently available), 67 (fuel used, reusability of), 76, 87, 93-94
rocks, 7, 8, **11**, 23, 29, 32, 33, 36, 40, **46**, 48, 54, 57, 58, 75
Roosa, Stu, *see also* Apollo 14, 36
Russia, *see also* Soviet, 5, 23-24, 26-31 (space race), 34, 37 (timeline), **44** (Luna 17 landing), 45 (Luna 24), **65** (Proton M), 66-68 (Federation), 70-71 (plans), **72** (female astronauts), 73-**74** (ISS toilet), 89

S

Saturn V, 16, 30, 61, 62, **63**, 65, 67, 77, 78, 95
Schmitt, Harrison "Jack," *see also* Apollo 17, 40, **43**
Shenzhou (spacecraft), 66
Schweickart, Rusty, *see also* Apollo 9, 31
Scott, David, *see also* Apollo 9 and Apollo 15, 22 (Apollo 9), 31, **38** (photo by)
sea, *see also* Sea of, 11, **18-19**, 24-27,31,40 (ages of), 44-46 (samples from), 47 (to find Apollo sites), 51-52, **55** (gravity map of), **56**, 58, 81 (magnetism of), 94
Sea of: Cleverness, 81; Clouds, 26, Crises, **10**, **18-19**, 31, **45**, **55**; Fertility, **18-19**, 44, **47**, Nectar, **18**, 40, 46 (age of), 47, 52; Rains, **18**, 40, **44** (Luna 17 in),46 (age of), **47**, **55-56** (relation to SPA), 58, Serenity, **18**, 42, 46 (age of), **47**, 56 , Tranquility, **18**, 21 (Apollo 11), 24-**25** (Ranger in), 27, 46 (age of), **47**, **55-56**
Shackleton Crater, 49
Shepard, Alan, *see also* Apollo 14, 36
Shoemaker, Eugene, 42, **50**
Shorty Crater, 42, **43**
SLS (rocket), 65, **66** (rocket of), 67-68, 69 (for module), 72 (for EM-1)
SMART-1 (spacecraft), 52
Smythe Sea, **19**
Soviet Union, *see also* Russia, 23, **27** (Luna 9), 44 (lunar exploration of)
South Pole-Atkin Basin, *see* SPA
Soyuz (spacecraft): Soyuz 1, 28, 31 (timeline); Soyuz TMA (capsule), 65, 68, 71;

Soyuz 5 (booster), 68
SPA Basin, **18-19**, 56 (antipode of Rains), **59**
Space Launch System, *see* SLS
space shuttle, 67, 71, 77, (fuel cells of), 90 (funding)
space station, *see also* ISS, 44 (first), 50, 65-66, 68 (Gateway), 70, **76** (solar arrays of), **77**-78 (batteries and radiators of), 80, **83** (plants on), 87, 89 (cost of)
SpaceX (company), *see also* Falcon, 61, **65** (Falcon Heavy), 66, 68, 70, 72 (flight of Maezawa)
Spudis, Paul, 51
Stafford, Tom, *see also* Apollo 10, 31
surface (of Moon), 8-10 (formation), 12 (view from), 17 (gravity), 18 (tour of), 19 (landing on), **20** (probes of), 21 (Apollo 11 landing on), 23-24, **25** (Ranger views of), 26-29 (hardness of), 30-33 (trash left on), 34, 36, 40, **41** (experiment on), 43 (gases on), 45, **48** (rifts under), 49, 51 (composition of), 54-**55** (olivine on), 57-58 (crashed on), **59** (photos of), 60, 63, 70, 72, 75, 81-**82** (lava tubes under), **85** (John Young on), 86, 88, 92-94
Surveyor (spacecraft), cover (in visor), **26** (image of Tycho), 27 (landings of), 31 (timeline), **34** (in painting and photo)
Swigert, Jack, *see also* Apollo 13, 35

T

temperatures, 50 (in craters), 53 (map of), 54 (in Cabeus Crater), 58, 67 (of fuel), 81 (underground), 88, 92 (extremes, average), 93-94
Tereshkova, Valentina, **72**
Theophilus Crater, 40
titanium, 29, 40 (where found), 45-46, **52** (abundance of), 80, 87, 90, 93, 94
Tsiolkovsky Crater, **19**, 38
Tycho Crater, **18**, **26** (by Surveyor 7), 27, 40, 57 (glow)

V

volcano/volcanic, 10-11, 23, 29 (rocks), **38** (in layers), 40, 46, 48, 50, 54, 93
Von Kármán Crater, **58-59**

W

water, 28, 46, 48-**50** (discovery of), 52-**53** (map of), 54, 59, 62, 67, 73-75, 77-78 (for electricity), 79-**80** (recycling), 83-84 (in hydroponics), 87-88, 93
White, Ed, *see also* Apollo 1, 28
Worden, Al, *see also* Apollo 15, 38

Y

Yang, Liu, **72**
Young, John, *see also* Apollo 10 and 16, 31, 40, **46** (Apollo 16), **85** (Apollo 16)
Yutu (rover), **58**

Z

Zero-g, *see* Freefall
Zond (spacecraft), 23 (timeline), **28** (turtles flown by), 30-31 (timeline), 34, 37 (timeline), 44, 71 (heat shield of), 73 (aerobraking by)